Stochastic Differential Equations for Chemical Transformations in White Noise Probability Space

Don Kulasiri

Stochastic Differential Equations for Chemical Transformations in White Noise Probability Space

Wick Products and Computations

Springer

Don Kulasiri
Centre for Advanced Computational
Solutions (C-fACS)
Lincoln University
Lincoln, Christchurch, New Zealand

ISBN 978-981-97-9391-4 ISBN 978-981-97-9392-1 (eBook)
https://doi.org/10.1007/978-981-97-9392-1

This Springer imprint is published by the registered company Springer Nature Singapore Pte Ltd.
The registered company address is: 152 Beach Road, #21-01/04 Gateway East, Singapore 189721, Singapore

To Sandhya for her love, support and care during the last 40 years journey together.

Preface

This monograph is an outcome of our curiosity to understand chemical transformations that occur ubiquitously all around us based on physical and mathematical paradigms. As our research spans from modeling intra-cellular processes especially in neurobiology to macroscale environmental processes such as contaminant transport in porous media during the last three decades, it is quite natural to delve into stochastic processes, stochastic differential equations (ordinary and partial), and thermodynamics related to dispersion and reactions to bring about deeper understanding and quantifying the complexity associated with chemical transformations. Our past research monographs and published works related to phenomena at macroscale and mesoscopic scales attest to our endeavor, and with our research monograph on chemical master equations for large biological networks in 2021 (see the references in Chap. 1) we attempt to solve the chemical master equations for a large number of biological reactions using the computational platform developed. Master equations render quintessential approaches to solve chemical reactions that occur in mesoscopic scales such as intra-cellular protein interactions. Low copy numbers, natural compartments, biological membranes acting as boundaries that exist within such environments, and random collisions of molecules would create stochasticity within these systems, so we conceptualize the interactions as noisy phenomena, hence the approaches like chemical master equations are used frequently. However, there is a long history over 100 years of chemical transformation in thermodynamics, and it is much more meaningful to incorporate the physics of chemical transformations into our modeling efforts.

In this monograph we attempt to incorporate the physics of equilibrium thermodynamics and nonequilibrium thermodynamics that have been developed to understand the self-organization that is observed in chemical reactions far from thermodynamic equilibrium. Most of the chemical reactions in life sciences happen in thermodynamically open systems with highly specialized porous boundaries, therefore we succinctly discuss the nonequilibrium thermodynamics, especially related to the entropy that plays a key role in understanding the stability of chemical reactions.

In terms of developing stochastic models, we make use of white noise probability space and the mathematics that has been developed during the last three to four decades using the Wick products and Malliavin calculus. This field is vast, and mathematics is elegant, unlike Ito calculus, we build the mathematical machinery using the rules similar to normal calculus. I develop the models using the Wick products and include some of the examples I have evaluated. To keep the monograph to a reasonable length, I have to compromise in the selection of examples and the mathematics, hopefully I have conveyed the gist sufficiently.

Chapter 1 introduces the thermodynamics associated with chemical reactions and we briefly summarize the results that are used in the later chapters. In Chap. 2 we present a brief background on Malliavin calculus and white noise analysis in a heuristic manner mentioning the key mathematics that are necessary for the computations. We also compute important functions and objects to show how they behave to build an intuitive feel that is necessary to understand the behaviors of models. We further delve into Wick algebra and show how to compute them to obtain approximations in Chap. 3. In Chap. 4 we discuss the Wick differential equations and discuss the positive noise functions that are used in the later chapters. In Chap. 5, we develop a numerical procedure to solve the Wick differential equations to obtain approximate solutions. We show the computations of such solutions, and emphasis is not on the rigor of numerical procedures but on the approximate behaviors. In Chap. 6, we discuss the deterministic models for chemical transformations and their stability, especially using perturbations and excess entropy production rate. Chapter 7 deals with the investigations of numerical solutions of the Wick differential equations, and in Chap. 8 we explore Kondepudi and Nelson model of chiral symmetry and develop a stochastic version of it to understand how system noise affects the model behavior around the bifurcation point. Hopefully, we have opened a novel avenue to explore the stochastic behavior of chemical reactions without resorting to expensive computations.

We provide Mathematica® codes in the book so that readers can use them to run their own computational experiments to consolidate their own understanding. In programming I only use basic Mathematica® language constructs and reader can see how the computations are carried out following the mathematics in the text. To keep this monograph brief, I have not included the outputs of the programs, but important program outputs are discussed in Chapters, but reader can run the programs and see the outputs that are not mentioned. I have tested these programs using the Mathematica® version 14 (Windows), and they run efficiently enough in modern desktop or laptop environments. Experienced Mathematica® programmers can obviously see ways to improve the efficiencies and elegance of codes but my intention is pedagogical.

As this monograph is interdisciplinary by nature, all the references that lead to the developments of physics and mathematics would be almost impossible to include at each chapter as these developments span over a century, and references lists will be extensive. Hence, I profusely apologize for many authors whose work I have not included. Instead, I am grateful for the authors of the following textbooks that I refer and follow quite often in this monograph. These books contain the references that are pertaining to the work I present here.

- Kondepudi, D. and Prigogine, I. (2014) Modern Thermodynamics: From Heat Engines to Dissipative Structures: Second Edition. Available at: https://doi.org/10.1002/9781118698723.
- Giulia Di Nunno, Bernt Øksendal and Frank Proske (2019) Malliavin Calculus for Lévy Processes with Applications to Finance. Berlin, Heidelberg: Springer-Verlag.
- Holden, H., Oksendal, B., Jan Uboe, and Zhang, T. (2010) Stochastic Partial Differential Equations: A Modeling White Noise Functional Approach. New York, NY: Springer. Available at: https://doi.org/10.1007/978-0-387-89488-1.

I acknowledge Lincoln University, Lincoln, Christchurch, New Zealand, for the facilitation of this work, and the Wolfson Centre for Mathematical Biology, the Mathematical Institute, the University of Oxford, UK for hosting me during my study leave in May/June in 2023. I also thank the editorial staff in SpringerNature helping in the publication of this monograph.

Lincoln, New Zealand
August 2024

Don Kulasiri

Contents

List of Figures

Chapter 1
Chemical Transformation in Far-from-Equilibrium Systems

1.1 Introduction

Chemical transformations are ubiquitous in the universe. We use the principles or the laws of thermodynamics that were developed over the last two centuries or more, to understand and calculate the entities involved in chemical transformations. (We use "transformations" and "reactions" synonymously in this book.) Since thermodynamics is a vast subject, we only succinctly discuss the concepts and equations that are essential to understand the subsequent chapters. The reader is referred to excellent references in this subject such as Kondepudi and Prigogine (2014), Prigogine and Van Rysselberghe (1963), Keizer (1979, 1987), and Nicolis and Prigogine (1971) for further study.

In thermodynamics, the nature of boundary around an identified system under study plays an important role in theoretical developments. If the boundary is impervious to the materials and energy flows, the systems is "isolated"; if the boundary allows only energy to transfer across, the system is "closed"; and if both the material and energy transfer in and out of the boundary, the system is "open". For most living systems in biology such as cells and organs have membranes often highly specialized through evolution, therefore the systems are naturally open. The open systems have special thermodynamic characteristics as we see later.

Thermodynamics is macroscopic science with deterministic variables in most systems. Hence, we focus on our attention first to macroscopic scale systems. Temperature, T, pressure, p, volume, V, and moles N_k of a chemical constituent k are a set of macroscopic variables, which we call the state of a system, and state variables are sufficient to define a system in thermodynamic equilibrium, in which the state variables are in steady state, and any perturbation from this state would cause the system to come back to the original equilibrium. Mathematically a thermodynamic equilibrium system must have definable variables that would reach stable extrema (minima/maxima). Hence stability is an important criterion around thermodynamic equilibrium. The first law of thermodynamics (conservation of energy, U) and the

D. Kulasiri, *Stochastic Differential Equations for Chemical Transformations in White Noise Probability Space*, https://doi.org/10.1007/978-981-97-9392-1_1

second law defining the concept of entropy, S, are the foundations of thermodynamics. Energy U of a system is expressed in terms of the rate of change in U that is the net inflow and out flow rates of energy across the boundary plus the rate of energy generation within the system due to chemical reactions This statement can equally be true for isolated, closed and open systems, and if the rate of change of U is zero, then the system is in a steady state with respect to energy, but we cannot say much about the thermodynamic equilibrium. For an observer measuring only U of a system in steady state would observe that purely from an energy perspective, time becomes irrelevant. S is created due to the irreversible processes such as diffusion in a system, and irreversible processes and their measure, S, will create a sense of time. The arrow of time has a definite direction of change in entropy, i.e., due to irreversible processes.

The problem of reversibility is an important issue in chemical transformation even though the plausibility of reversibility exists in any irreversible reaction but in many cases, it is rarely seen in nature. For example, sugar dissolves in water but sweet water never produces sugar back in solid form under the same conditions. This is quite different in theories of mechanics: Newton laws for macroscopic systems and quantum mechanics for microscopic phenomena; both laws are time reversible (Prigogine & Stengers, 2017). This observation can be interpreted differently and discussion on this topic is beyond the scope of this book, but the reader is referred to Prigogine's work on this subject. Because the theory of thermodynamics was first developed for equilibrium systems, we discuss equilibrium related to chemical transformation, to narrow down the scope.

1.2 The Modern Formulation of Entropy

1.2.1 Introduction

Prigogine and Defay (1967), combining the advances made by De Donder (Kondepudi & Prigogine, 2014), developed the modern approach to understand irreversibility in chemical transformations using the second law of thermodynamics and formulated the ways to calculate entropy due to chemical reactions The corner stone of the modern formalism is the local equilibrium assumption. The macroscopic thermodynamic variables, temperature, pressure, mole concentration and internal energy are assumed to exist within an infinitesimally small volume of a system (ΔV), and within ΔV, it is assumed that the thermodynamic equilibrium locally exists. Hence, we can define densities for extensive variables of the system. For example, temperature, T, in a macroscopic system is a concept that measures the "hotness" of the system. Higher T, the system contains more "heat", which is a form of energy. T can be defined according to the Maxwell distribution of velocities of colliding molecules. According to statistical mechanics (Kondepudi & Prigogine, 2014), if a system of molecules is in equilibrium, the velocity distribution of molecules follows the Maxwell distribution.

According to molecular dynamics simulations, the Maxmillian velocity distribution is reached within less than 10^{-9} s in a system with gas at 1 atm pressure after perturbation. This means that, for very many chemically reacting systems, the collisions are rapid enough to define a temperature at any elemental volume ΔV. It can also be shown that molar density is a meaningful quantity within a small volume of the size of 10^{-15} L (Kondepudi & Prigogine, 2014). This means that if the characteristic size of a cube is 1 μm, then the molar density within 1 $(\mu m)^3$ volume is uniform and in equilibrium. So, the local equilibrium assumption can be valid for a very small systems in biology; hence, we can use the formalism that has been adopted by Prigogine and his colleagues.

1.2.2 Chemically Reactive Systems in Thermodynamic Equilibrium

Before proceeding to nonequilibrium systems, we briefly introduce the concepts and the development of equations that would help us to understand the theoretical developments in the later chapters. Reader may refer to the excellent textbooks and the references for further study (Keizer, 1987; Prigogine & Defay, 1967; Ross, 2008).

The state function called enthalpy, H, defined as

$$H \equiv U + pV \tag{1.1}$$

where U is the internal energy, p is the pressure and V is the volume, is used in calculating the heat released by a chemical reaction. The enthalpy change in a chemical reaction under constant pressure is called the enthalpy of reaction, ΔH_{r}. ΔH_{r} is positive for endothermic reactions and is negative for exothermic reactions. If a chemical reaction is a combined one with two steps, ΔH_{r} is the sum of individual ΔH_{r} of each reaction. This is called Hess's law and can be used to calculate ΔH_{r} of many reactions. As we are dealing with constant pressure systems, the energy conservation is expressed using enthalpies.

Historically chemical reactions of all sorts were the domain of chemists, and they developed the branch of empirical science of chemical kinetics. Chemical kinetics deals with how reactions occur, reaction rates and stoichiometry of reactions. Kinetics is based on kinetics laws such as mass action and Micheal-Menten that are based on extensive experimental studies. With the advent of Newtonian mechanics along with concept of force, it was a natural to seek a similar "force" that drives a chemical reaction. The idea of "affinity" of a chemical reaction was prevalent among chemists as the "force" that caused chemical reactions. The concept of chemical potential proposed by J. W. Gibbs was adapted to chemical reactions, and De Donder's formulation of chemical affinity was based on the idea of chemical potential. Based on Gibbs formulation, the change in energy $\mathrm{d}U$ due to a chemical reaction can be expressed as follows:

$$\mathrm{d}U = T\mathrm{d}S - pV + \sum_{j}^{n} \mu_j \mathrm{d}N_j, \tag{1.2}$$

where T, S, p, V, and N_j are temperature, entropy, pressure, volume, and molar amounts of the species, respectively. The coefficients μ_j are called the chemical potentials. It is assumed that the system is in equilibrium and the chemical transformations between equilibrium states are reversible, then only the change of heat,

$$\mathrm{d}Q = T\mathrm{d}S.$$

The following relationships are implied in Eq. (1.2):

$$\left(\frac{\partial U}{\partial S}\right)_{V,N_j} = T; \quad \left(\frac{\partial U}{\partial V}\right)_{S,N_j} = -p; \quad \text{and} \quad \left(\frac{\partial U}{\partial N_k}\right)_{S,V,N_{k\neq j}} = \mu_j.$$

For a system, where the boundaries are porous with respect to reversible exchange of matter and energy, we can express the change in the molar amounts due to the reaction k among n reactions as,

$$\mathrm{d}N_k = \mathrm{d_i}N_k + \mathrm{d_e}N_k, \tag{1.3}$$

where $\mathrm{d_i}N_k$ is the change due to irreversible chemical reactions and $\mathrm{d_e}N_k$ is the reversible change due to the exchange of matter with the exterior through the boundary. Equation (1.2) can be written as,

$$\mathrm{d_e}S = \frac{\mathrm{d}U + p\mathrm{d}V}{T} - \frac{\sum_1^n \mu_k \mathrm{d_e}N_k}{T}. \tag{1.4}$$

De Donder associated the second term on the right side of Eq. (1.4) as,

$$\mathrm{d_i}S = -\frac{\sum_1^n \mu_k \mathrm{d_e}N_k}{T}.$$

For a closed system, $\mathrm{d_e}N_k = 0$, by definition, then

$$\mathrm{d_i}S = -\frac{\sum_1^n \mu_k \mathrm{d_e}N_k}{T} > 0. \tag{1.5}$$

The rate of entropy production is positive if any of the reactions are irreversible. For closed systems, De Donder's Eq. (1.5) relates the entropy production rate to the kinetics of chemical reactions. In other words, a system away from chemical equilibrium, the chemical reactions will irreversibly drive the system toward equilibrium in such a way that the second law of thermodynamics is adhered to through Eq. (1.5).

The concept of affinity for closed systems as discussed by Kondepudi and Prigogine (2014) is summarized here without giving any details. For a general chemical reaction of the form,

$$a_1A_1 + a_2A_2 + \cdots + a_nA_n \rightleftarrows b_1B_1 + b_2B_2 + \cdots + b_mB_m,$$

where $a_i\,(i = 1\ldots n)$ and $b_j\,(j = 1\ldots m)$ are stoichiometric coefficients corresponding to A_i s (reactants) and B_i s (products), respectively. In this reaction, only independent variable is the extent of reaction (ξ) as defined by,

$$\frac{\mathrm{d}N_{A_1}}{-a_1} = \frac{\mathrm{d}N_{A_2}}{-a_2} = \ldots = \frac{\mathrm{d}N_{A_n}}{-a_n} = \frac{\mathrm{d}N_{B_1}}{b_1} = \frac{\mathrm{d}N_{B_2}}{b_2} = \ldots = \frac{\mathrm{d}N_{B_m}}{b_m} = \mathrm{d}\xi,$$

where N_{A_i} and N_{B_j} are the molar amounts of reactants and products, respectively. Once ξ is calculated using kinetics combined with mass conservation law, we compute the molar concentration of each species.

1.2.3 Thermodynamic Flows and Fluxes

The second law of thermodynamics that defines entropy is a vast subject and we do not attempt to discuss it in detail, but the modern formulation of entropy would be useful for our work in later chapters. The second law can be stated as: the sum of entropy changes of a system and its exterior can never decrease, i.e., the entropy of an isolated system always approaches a maximum in reaching the thermodynamic equilibrium.

Based on the local equilibrium assumption, the changes of entropy of a system can be expressed as,

$$\mathrm{d}S = \mathrm{d_e}S + \mathrm{d_i}S \tag{1.6}$$

where $\mathrm{d_e}S$ is the change in entropy due to the exchange of matter and energy across the boundary and $\mathrm{d_i}S$ is the change in entropy due to the irreversible processes happening within the system. Analogous to mechanics, we can think of thermodynamics flows such as heat are driven by thermodynamic forces such as temperature differences, and this process is irreversible. The change in entropy inside the system due to irreversible processes can be expressed as,

$$\mathrm{d_i}S = F\,\mathrm{d}X$$

where F is the thermodynamic force and X is the flow. For a system either isolated, open or closed, a modern statement of the second law can be expressed for a system containing k number of irreversible processes

$$\frac{dS_i}{dt} = \sum_k F_k \frac{dX_k}{dt} \geq 0.$$

This statement is valid for each subsystem of a system individually. In other words, the total entropy production rate is nonnegative with each subsystem having a nonnegative entropy production rate. If we take Eq. (1.6), we can have situations of $dS < 0$ if $d_eS < 0$ but absolute value of which is numerically larger than d_iS, which is always nonnegative. This interpretation is useful in living systems where disorder is decreased with self-organizing behaviors.

The concept of affinity originally defined by De Donder crystallizes the idea of thermodynamic force in chemical reactions. The affinity A of the general reaction given can be defined as,

$$\mathrm{A} \equiv \sum_{k=1}^{n} a_k \mu_{A_k} - \sum_{k=1}^{m} b_k \mu_{B_k} \tag{1.7}$$

where μ_{A_k} and μ_{B_k} are the chemical potentials of the reacting species A_k and the product species B_k, respectively. Given this definition, we can express the rate of entropy production in a closed system when several reactions occur simultaneously in terms of an affinity A_k and a degree of advancement ξ_k:

$$\frac{d_iS}{dt} = \sum_k \frac{\mathrm{A}_k}{T} \frac{d\xi_k}{dt} \geq 0. \tag{1.8}$$

At thermodynamic equilibrium, the affinity A and the velocity $\frac{d\xi}{dt}$ of each reaction are zero. It should be mentioned as noted by Kondepudi and Prigogine (2014) Gibbs free energy of reversible chemical reactions at equilibrium is numerically computed by multiplying affinity by -1:

> …but there is a fundamental conceptual difference between the two: affinity is a concept that relates to irreversible chemical reactions to entropy, whereas the Gibbs free energy is primarily used in connection with equilibrium states and reversible processes.

Affinity just like entropy is a state function defined by chemical potentials. For a closed system, affinity is a function of V, T, and ξ, and the initial values of molar amounts. There are several properties of affinity that can be useful in applications:

(1) If affinity of a reaction is positive, then the reaction proceeds to the right;
(2) If affinity is negative, then it proceeds to the left;
(3) The affinity of a reaction made of several sequential reactions is the sum of the affinities of individual reactions; and

(4) The rate of entropy production of each step in coupled reactions are positive. (See Kondepudi and Prigogine (2014) for details.)

Further, it can be shown that the process of diffusion of molecules from a region of higher concentration to lower concentration is driven by the thermodynamic force of the gradient of chemical potential, not by the gradient of concentration as mostly stated in textbooks. (See Kondepudi and Prigogine (2014).)

We define reaction rate and reaction velocity using A and ξ. Here the number of reactive events per second per unit volume is defined as the reaction rate. As chemical reactions depend on collisions and a very small number of collisions produce a reaction, the rates of reactions are therefore proportional to the concentrations of reactants and products. For example, a reaction of the form,

$$A + B \rightleftarrows C + D,$$

the forward reaction rate

$$R_\mathrm{f} = k_\mathrm{f}[A][B],$$

and the reverse rate

$$R_\mathrm{r} = k_\mathrm{r}[C][D],$$

where k_f and k_r are the rate coefficients. We define the velocity of the reaction as the rate of net conversions of the reactants to products per unit volume. ([ ] indicates the concentrations of the species.) Reaction velocity, ν, can therefore be written as,

$$\nu = R_\mathrm{f} - R_\mathrm{r} = k_\mathrm{f}[A][B] - k_\mathrm{r}[C][D], \tag{1.9}$$

and υ is the rate of change of ξ (extent of reaction) per unit volume:

$$\nu = \frac{1}{V}\frac{\mathrm{d}\xi}{\mathrm{d}t} = R_\mathrm{f} - R_\mathrm{r}. \tag{1.10}$$

Equation (1.10) is useful in developing reaction kinetic models for reaction systems as often seen in applications. A and ξ can be written as,

$$\mathrm{A} = \mu_A + \mu_B - \mu_C - \mu_D, \text{ and}$$

$$\mathrm{d}\xi = \frac{\mathrm{d}N_A}{-1} = \frac{\mathrm{d}N_B}{-1} = \frac{\mathrm{d}N_C}{1} = \frac{\mathrm{d}N_D}{1}.$$

We can write the rate of entropy production due to the reaction as,

$$\frac{\mathrm{d}_i S}{\mathrm{d}t} = \left(\frac{\mathrm{A}}{T}\right)\frac{\mathrm{d}\xi}{\mathrm{d}t},$$

$$= -\frac{(\mu_A + \mu_B - \mu_C - \mu_D)}{T}\frac{\mathrm{d}N_{A\,\mathrm{or}\,B}}{\mathrm{d}t}, \text{ or}$$

$$= \frac{(\mu_A + \mu_B - \mu_C - \mu_D)}{T}\frac{\mathrm{d}N_{C\,\mathrm{or}\,D}}{\mathrm{d}t}.$$

Therefore, for a closed system, the thermodynamic force is $\left(\frac{A}{T}\right)$ and the thermodynamic flow is $\frac{\mathrm{d}\xi}{\mathrm{d}t}$; the entropy production due to the reaction is always positive if the reaction velocity is positive, and the entropy reaches a maximum when the system approaches thermodynamic equilibrium.

It should be noted that the reaction rates are not as in Eq. (1.9)(mass action law) but can be in the form kA^aB^b for the forward rate for example where a and b are stoichiometric coefficients of the reaction,

$$A + B \rightleftarrows C + D.$$

Further k may depend on T as given by,

$$k = k_0 \mathrm{e}^{-(E_\mathrm{a}/RT)},$$

where k_0 is pre-exponential factor and E_a is the activation energy. This is called the Arrhenius equation and is accurate when T varies within small limits. Otherwise, a more accurate theory based on transition states can be used (Kondepudi & Prigogine, 2014).

Some chemical reactions are composite reactions with multiple steps. A step that is not a combined reaction is called an elementary reaction. We often use elementary reactions in this book. Chemical reactions that are in thermodynamic equilibrium adhere to the principle of detailed balance, i.e., every chemical transformation is balanced by its exact opposite or reverse. This means that $(k_\mathrm{f}/k_\mathrm{r})$ is invariant with respect to kinetic mechanism involved in forward and reverse reactions. This principle is violated when the reaction occurs far from thermodynamic equilibrium.

1.3 Fluctuations and Stability in Chemically Reactive Systems

At equilibrium, the entropy is maximum for an isolated system. This means that any fluctuation should reduce the entropy, and then the irreversible reactions that produce entropy should bring the system back to the equilibrium. Therefore, equilibrium state is stable to perturbations that result in decrease in entropy We can write the entropy of the system as,

$$S = S_{\mathrm{eq}} + \delta S + \frac{1}{2}\delta^2 S + \cdots + \cdots,$$

using Taylor expansion. The first order term δS depends on δV, δT or δP; and the second order term $\delta^2 S$ depends on δV^2, δT^2, etc., i.e., the perturbations of second order. For an isolated chemical reaction, fluctuations in the molar amounts are stable if,

$$S - S_{\text{eq}} = \Delta S = \delta S + \frac{1}{2}\delta^2 S < 0$$
$$= \left(\frac{\partial S}{\partial \xi}\right)_{U,V} \delta\xi + \frac{1}{2}\left(\frac{\partial^2 S}{\partial \xi^2}\right)^2_{U,V} (\delta\xi)^2 < 0.$$

At equilibrium,

$$\left(\frac{\partial S}{\partial \xi}\right)_{U,V} = \frac{\mathrm{A}}{T},$$

$$\therefore \ \Delta S = \left(\frac{\mathrm{A}}{T}\right)_{\text{eq}} \delta\xi + \frac{1}{2T}\left(\frac{\partial \mathrm{A}}{\partial \xi}\right)_{\text{eq}} (\delta\xi)^2. \tag{1.11}$$

In Eq. (1.11), the first term on the right-hand side (rhs) is zero as $\mathrm{A} = 0$ at equilibrium, hence, $\left(\frac{\partial \mathrm{A}}{\partial \xi}\right)_{\text{eq}} < 0$, given $T > 0$.

The partial derivative of affinity of chemical reaction with respect to the extent of reaction should be negative for the system to be stable, i.e., for the entropy to reach its maximum when perturbed. When a number of reactions, r, are involved, we can generalize above condition for the following statement:

$$\frac{1}{2}\delta^2 S = \sum_{i,j}^{r} \frac{1}{2T}\left(\frac{\partial \mathrm{A}_i}{\partial \xi_j}\right)_{\text{eq}} \delta\xi_i \delta\xi_j < 0. \tag{1.12}$$

Equation (1.12) can be written in terms of chemical potentials μ_k and molar amounts N_k:

$$\Delta_i S = -\sum_{i,j}^{r} \frac{1}{T}\left(\frac{\partial \mu_i}{\partial N_j}\right)_{\text{eq}} (\delta N_i \delta N_j)/2 < 0.$$

This relationship is independent of boundary conditions of the system and can be applied to fluctuations due to chemical reactions as well. Duhem–Jouget theorem states that if a system is stable due to diffusion in equilibrium, then it is stable to chemical reactions. When we incorporate the fluctuations related to temperature, volume and chemical potential, then the general condition for the stability can be expressed as (Kondepudi & Prigogine, 2014),

$$\delta^2 S = -\frac{C_v}{T^2}(\delta T)^2 - \frac{1}{T\mathrm{K}_T}\frac{(\delta V)^2}{V} - \sum_{i,j}^{r}\frac{1}{T}\left(\frac{\partial \mu_i/T}{\partial N_j}\right)_{\mathrm{eq}}(\delta N_i \delta N_j) < 0,$$

where C_v is the heat capacity of the system with arbitrary volume V and K_T is the isothermal compressibility,

$$\mathrm{K}_T = -\left(\frac{1}{V}\right)\left(\frac{\partial V}{\partial P}\right).$$

1.3.1 Stability in Nonlinear Thermodynamic Systems with Chemical Reactions

Most of living systems in biology are thermodynamically open systems, therefore, we need to compute the entropy flows associated with the exchange of matter and energy. Let us take the example of the following chemical reaction in an open system that depicts a monomolecular transformation such as isomerization:

$$A \rightleftarrows X \rightleftarrows B,$$

and we designate (A, X) reaction as reaction 1 and (X, B) reaction as reaction 2. As we invoke the local equilibrium assumption, we express all the extensive variables such as molar amounts as densities, i.e., dividing quantities by the volume. We can write the entropy production per unit volume (σ) in isothermal conditions as,

$$\sigma = \frac{\mathrm{A}_1}{T}\nu_1 + \frac{\mathrm{A}_2}{T}\nu_2 > 0,$$

where A_k and ν_k $(k = 1, 2)$ are affinities and the velocities of two reactions, respectively. We can also write,

$$\nu_k = R_{k\mathrm{f}} - R_{k\mathrm{r}}, \text{ and}$$

$$\mathrm{A}_k = RT\,\ln\left(\frac{R_{k\mathrm{f}}}{R_{k\mathrm{r}}}\right), \quad \text{for } k = 1, 2.$$

As in living systems, to keep the system at a constant temperature, the heat of reaction should be compensated by the net heat flow of the system. To keep the system at stationary state,

$$\frac{\mathrm{d}S}{\mathrm{d}t} = \frac{\mathrm{d_e}S}{\mathrm{d}t} + \frac{\mathrm{d_i}S}{\mathrm{d}t} = 0,$$

where $\frac{d_i S}{dt} = \int \sigma \, dV$.

This is only possible when the entropy exchange with exterior is negative, i.e.,

$$\left(\frac{d_e S}{dt} < 0 \right).$$

It can be shown that when the local equilibrium condition is invoked, the entropy current (flow) can be written as, (See Kondepudi and Prigogine (2014) for details.)

$$J_s = \frac{J_q}{T} + \sum_k S_k J_k,$$

where $S_k = \left(\frac{\partial s}{\partial n_k} \right)_T$, partial molar entropy (entropy per mol) and J_q is the heat flow. Entropy density, s, and molar density, n, are per unit volume. With this notation, it can be shown that,

$$\frac{d_e S}{dt} = -\frac{1}{T} \oint J_q \cdot da + \oint S_A J_A \cdot da + \oint S_B J_B \cdot da < 0, \tag{1.13}$$

where a is the surface area vector enclosing the system. Rhs of Eq. (1.13) is negative when the system is in nonequilibrium steady states (NESS), and Eq. (1.13) can be applied to chemically reactive systems.

In the linear regime of the nonequilibrium thermodynamic states, the reciprocal relationships developed by Onsager are valid, and it can be shown that the total entropy production in a system with the flows of energy and matter across the boundary is minimum at the NESS (Glansdorff et al., 1974), i.e.,

$$\frac{d_i S}{dt} = \int \sigma \, dV$$

is a minimum.

The rate of entropy change can be thought of as a potential (P) and can be expressed as follows:

$$P \equiv \frac{d_i S}{dt} = \int \sum_k F_k J_k dV,$$

where F_k, J_k are coupled forces and flows, respectively. For simplicity, let $k = 2$, and we have

$$P = \int (F_1 J_1 + F_2 J_2) dV.$$

Onsager's linear phenomenological laws with L_{ij}, phenomenological coefficients, give,

$$J_1 = L_{11}F_1 + L_{12}F_2$$

and

$$J_2 = L_{21}F_1 + L_{22}F_2.$$

With Onsager's reciprocal relations, $L_{12} = L_{21}$,

$$P = \int (L_{11}F_1^2 + 2L_{12}F_1F_2 + L_{22}F_2^2)\mathrm{d}V.$$

For a fixed F_1,

$$\frac{\partial P}{\partial F_2} = \int 2(L_{22}F_2 + L_{12}F_1)\mathrm{d}V = \int 2(J_2)\mathrm{d}V,$$

and at a minimum with respect to F_2,

$$\frac{\partial P}{\partial F_2} = 0, \quad \text{i.e.,} \quad J_2 = 0.$$

Hence $P(\equiv \frac{\mathrm{d}_i S}{\mathrm{d}t})$ is minimized when the flow related to the unconstrained force, F_2, is diminished to zero. So, at the NESS, the system has a minimum entropy production when unconstrained flow is zero. In general, P is minimum when the flows J_k related to the unconstrained forces, F_k, are zero. P can be thought therefore as a potential that drives the system away from the NESS. In other words, the rate of entropy production due to the reactions within the system pushes the system away from the NESS if the system is already in the NESS. The stability of NESS and equilibrium state can be understood by using P and $\frac{\mathrm{d}P}{\mathrm{d}t}$ as the Lyapunov conditions for the stability (Nicolis & Prigogine, 1979):

$$P > 0 \text{ and } \frac{\mathrm{d}P}{\mathrm{d}t} < 0. \tag{1.14}$$

If chemical reactions are closer to equilibrium, we can use the Onsager's reciprocal relationships to derive the relationships between the entropy production per unit volume, σ, and the thermodynamic forces and flows. But chemical reactions are often operate in nonlinear regime far from thermodynamic equilibrium in open systems.

1.3.2 Stability in Far-from-Equilibrium Systems

As previously mentioned, open systems often operate far from thermodynamic equilibrium, i.e., the thermodynamic flows are not linear functions of thermodynamic forces. Due to large affinities involved in chemical reactions, they easily drift into nonlinear region in open systems. (See Kondepudi and Prigogine (2014).) The stability of far-from-equilibrium states can easily be lost due to irreversible processes and the inherent fluctuations in the system. The fluctuations therefore play a major role in determining the potential new states that available to the system. These new states may produce highly self-organized systems and are called dissipative structures that are highly unpredictable, and the external and internal fluctuations dominate the behaviors of the system. In other words, probability laws become a part of natural laws, and it is impossible to disregard fluctuations in these phenomena. Bohm (2004) discusses at length the role the chance plays in physics. However, before discussing the fluctuation driven systems, we can still discuss far-from-equilibrium systems using the entropy production at the NESS.

There exists an extensive body of literature on the dissipative structures in chemical transformations, (See, for example, Prigogine and Nicolis (1967) and Nicolis and Prigogine (1971, 1979).) hence we very briefly state pertinent results that are relevant to the scope of this monograph. There are however general properties of entropy production that are still valid for the systems that evolve into different states producing self-organized states, especially around NESSs. Infinitesimal changes in the thermodynamic forces δF_k and flows δJ_k due to kth chemical reaction can be used to derive P, the rate of entropy production around a NESS:

$$P \equiv \frac{\mathrm{d_i}S}{\mathrm{d}t} = \int \sigma \,\mathrm{d}V = \int \sum_k F_k J_k \mathrm{d}V.$$

Then,

$$\frac{\mathrm{d}P}{\mathrm{d}t} = \int \left(\sum_k \frac{\mathrm{d}F_k}{\mathrm{d}t} J_k \right) \mathrm{d}V + \int \left(\sum_k F_k \frac{\mathrm{d}J_k}{\mathrm{d}t} \right) \mathrm{d}V, \tag{1.15}$$

$$\equiv \frac{\mathrm{d}_F P}{\mathrm{d}t} + \frac{\mathrm{d}_J P}{\mathrm{d}t}. \tag{1.16}$$

The first term on the rhs of Eqs. (1.15) and (1.16) is the rate of change of P due to the changes in the forces and the second term is due to the rate of changes of flows. It can be shown that in the linear regime,

$$\frac{\mathrm{d}_F P}{\mathrm{d}t} = \frac{\mathrm{d}_J P}{\mathrm{d}t},$$

and in the nonlinear regime,

$$\frac{d_F P}{dt} \leq 0,$$

and the equality holds, i.e.,

$$\frac{d_F P}{dt} = 0, \quad \text{at NESS.}$$

It can be shown that around a NESS, the second variation of entropy,

$$\frac{1}{2}\delta^2 S < 0,$$

where S is the total entropy of the system. This leads to the following result: A NESS is stable if

$$\frac{d\left(\frac{1}{2}\delta^2 S\right)}{dt} = \int \sum_k \delta F_k \delta J_k dV > 0.$$

However, $\sum_k \delta F_k \delta J_k < 0$ is not a sufficient condition for instability but a necessary one. For both equilibrium and nonequilibrium conditions, $\delta^2 S < 0$, but the stability of NESS requires,

$$\frac{d\left(\frac{1}{2}\delta^2 S\right)}{dt} > 0.$$

Hence $-\frac{1}{2}\delta^2 S$ is a Lyapunov functional. $\frac{d(\frac{1}{2}\delta^2 S)}{dt}$ is also called the excess entropy production in literature.

If $\boldsymbol{X}$ is the state of the system and $\Delta \boldsymbol{X}$ is the distance from equilibrium ($\Delta \boldsymbol{X} = \boldsymbol{0}$ at NESS), when $\boldsymbol{X}$ is sufficiently small, small perturbations would restore the system to $\boldsymbol{X}_s$ ($\boldsymbol{X}$ at NESS). $\boldsymbol{X}_s$ could move along the thermodynamic branch depending on the control parameters of the system. If the Lyapunov conditions are violated, the thermodynamic branch could be unstable, and the system may transit to a new branch, which may be a self-organized structure.

1.4 Importance of Fluctuations and Stochastic Processes Based on White Noise Analysis

Chemical reacting systems within cells for example operate in far-from equilibrium; hence dissipative structures may play important roles along with stability, bifurcations and state transitions of these reactions. Adding to this complexity low molecular numbers and collisions in crowded environments would create rather noisy environment in the background. Therefore, stochastic mathematical approaches would provide us with unique insights of behavior of biomolecular "reactions" if we could formulate the models of the systems (Kulasiri, 2011; Kulasiri & Kosarwal, 2021; Xie & Kulasiri, 2010). In addition, white noise formalism would be ideally suited to these problems at least theoretically (Di Nunno et al., 2019; Holden & Øksendal, 2022; Øksendal, 2008) hence the emphasis of developing appropriate models based on white noise stochastic differential equations in this monograph. However, the mathematical machinery is sophisticated and highly technical, and it is a challenge to make them more applicable to chemical transformations. The following chapter provides relevant mathematics prior to developing models and their numerical solutions in later chapters.

References

Bohm, D. (2004). *Causality and chance in modern physics*. https://doi.org/10.4324/9780203201107

Di Nunno, G., Øksendal, B., & Proske, F. (2019). *Malliavin calculus for Lévy processes with applications to finance*. Springer-Verlag.

Glansdorff, P., Nicolis, G., & Prigogine, I. (1974). The thermodynamic stability theory of non-equilibrium states. *Proceedings of the National Academy of Sciences, 71*(1). https://doi.org/10.1073/pnas.71.1.197

Holden, H., & Øksendal, B. (2022). A white noise approach to stochastic differential equations driven by Wiener and Poisson processes. In *Nonlinear theory of generalized functions*. https://doi.org/10.1201/9780203745458-28

Keizer, J. (1979). Nonequilibrium thermodynamics and the stability of states far from equilibrium. *Accounts of Chemical Research, 12*(7), 243–249. https://doi.org/10.1021/ar50139a004

Keizer, J. (1987). *Statistical thermodynamics of nonequilibrium processes*. Springer New York. https://doi.org/10.1007/978-1-4612-1054-2

Kondepudi, D., & Prigogine, I. (2014). *Modern thermodynamics: From heat engines to dissipative structures* (2nd ed.). https://doi.org/10.1002/9781118698723

Kulasiri, D. (2011). Computing molecular fluctuations in biochemical reaction systems based on a mechanistic, statistical theory of irreversible processes. *Methods in Enzymology*. https://doi.org/10.1016/B978-0-12-381270-4.00009-3

Kulasiri, D., & Kosarwal, R. (2021). *Chemical master equation for large biological networks: State-space expansion methods using AI*. https://doi.org/10.1007/978-981-16-5351-3

Nicolis, G., & Prigogine, I. (1971). Fluctuations in nonequilibrium systems. *Proceedings of the National Academy of Sciences, 68*(9), 2102–2107. https://doi.org/10.1073/pnas.68.9.2102

Nicolis, G., & Prigogine, I. (1979). Irreversible processes at nonequilibrium steady states and Lyapounov functions. *Proceedings of the National Academy of Sciences of the United States of America, 76*(12). https://doi.org/10.1073/pnas.76.12.6060

Øksendal, B. (2008). Stochastic partial differential equations driven by multi-parameter white noise of Lévy processes. *Quarterly of Applied Mathematics, 66*(3). https://doi.org/10.1090/s0033-569x-08-01090-5

Prigogine, I., & Defay, R. (1967). *Chemical thermodynamics* (4th ed.). Longmans.

Prigogine, I., & Nicolis, G. (1967). On symmetry-breaking instabilities in dissipative systems. *The Journal of Chemical Physics, 46*(9). https://doi.org/10.1063/1.1841255

Prigogine, I., & Stengers, I. (2017). *Order out of chaos man's new dialogue with nature*. Verso.

Prigogine, I., & Van Rysselberghe, P. (1963). Introduction to thermodynamics of irreversible processes. *Journal of The Electrochemical Society, 110*(4). https://doi.org/10.1149/1.2425756

Ross, J. (2008). *Thermodynamics and fluctuations far from equilibrium*. Springer Berlin Heidelberg. https://doi.org/10.1007/978-3-540-74555-6

Xie, Z., & Kulasiri, D. (2010). On exploring effects of molecular noise in a simple viral infection model. *International Journal of Biomathematics, 3*(1). https://doi.org/10.1142/S1793524510000891

Chapter 2
A Brief Background on Malliavin Calculus and White Noise Analysis

2.1 Introduction

Functional analysis provides us with a rich repertoire of algebraic structures as we see later, which could be used to develop models of complex systems having inherent probabilistic aspects to them. As often happens in physical sciences, the development of concepts related to the systems at hand are crucial to understanding the behaviors of systems. Mathematical structures that are employed in the concepts are as important as the concepts themselves, and in many cases, mathematics inspires conceptual developments. (Refer Kulasiri and Kosarwal (2021), Holden et al. (1996), Øksendal (1998), Di Nunno et al. (2019), Holden and Øksendal (2022), in the references.)

In this chapter, we are only concerned with the processes in continuous time, hence our central mathematical object is the Wiener process, the mathematical representation of Brownian motion developed by Nobert Wiener in 1931. In mathematical literature, $B(t, \omega)$ and $W(t, \omega)$ are both used as symbols for Brownian motion and Wiener process. We use both these notations in this book and should not create any confusion. But when we write white noise process, we use the lower case $w(t, \omega)$ and then again it will be clear within the context. There are excellent publications related to this subject, but we follow two books very closely: Di Nunno et al. (2019) and Holden et al. (2010) including their notation; the reason is that the mathematicians who wrote these books are pioneers in developing the theory of stochastic differential equations that we attempt to apply here. This chapter is not an exhaustive treatment neither of Malliavin calculus nor of white noise analysis. Readers are encouraged to consult these texts for further details.

As we encounter multi-index notation often in the white noise analysis, it is important to be familiar with it, and we attempt to interpret mathematical equations in analogues settings so that we can intuit the mathematics. Intuition often helps in developing models to understand the complexity of systems.

D. Kulasiri, *Stochastic Differential Equations for Chemical Transformations in White Noise Probability Space*, https://doi.org/10.1007/978-981-97-9392-1_2

In stochastic analysis, the Wiener–Itô chaos expansion play a pivotal role, which represents the square integrable random variables in terms of a Hilbert-space orthogonal sum due to Wiener (1938) and Itô subsequently (Ito, 1951) developed the iterated Itô integrals in the Wiener space. We discuss these developments in a heuristic manner and for more rigorous treatment Di Nunno et al. (2019) and Holden et al. (2010) are good references as previously mentioned.

We define

$$W = W(t) = W(t, \omega),\ \ t \in [0, T],\ \ \omega \in \Omega\ (T > 0),$$

a one-dimensional Wiener process (Brownian motion) on a probability space $(\Omega, \mathcal{F}, P)$, Wiener space with $W(0) = 0$. $\mathbb{F} = \{\mathcal{F}_t, t \in [0, T]\}$ is the filtration, $0 \le s \le t$, also consisting of all the P-zero measure events, and $\mathcal{F}_t$ is the σ-algebra generated by $W(s)$.

2.2 Iterated *n*-Fold Itô Integrals and Their Properties

The n-fold Itô integrals play a central role in the development of white noise analysis and modeling. These integrals sum the contributions of one-dimensional Itô integral at a particular time over a larger interval of time. If a deterministic function of time is used as a diffusion term in an Itô diffusion, the integration of which up to that time can be interpreted as the effects of the fluctuations within that time interval. Since chemical reactions are a result of molecular collisions, it can be hypothesized that the Brownian motion is a good model to reflect the nature of colliding molecules. The n-fold Itô integrals are sophisticated mathematical objects we can use to model the diffusive and colliding nature of molecules that produce chemical reactions.

Let us define f as a deterministic function on the set,

$$S_n = \left\{(t_1, \dots, t_n) \in [0, T]^n;\ \ 0 \le t_1 \le t_2 \le \dots \le t_n \le T\right\}, \tag{2.1}$$

with the norm defined on $L^2(S_n)$ as,

$$\|f\|^2_{L^2(S_n)} := \int\limits_{S_n} f^2(t_1, t_2, \dots, t_n)\mathrm{d}t_1\mathrm{d}t_2 \dots \mathrm{d}t_n < \infty. \tag{2.2}$$

Then the n-fold iterated Itô integral is defined as,

$$J_n(f) := \int\limits_0^T \int\limits_0^{t_n} \dots \int\limits_0^{t_3} \int\limits_0^{t_2} f(t_1, \dots, t_n)\mathrm{d}W(t_1)\mathrm{d}W(t_2) \dots \mathrm{d}W(t_{n-1})\mathrm{d}W(t_n).$$

Because the integrand

$$\int_0^{t_i}\int_0^{t_{i-1}}\cdots\int_0^{t_3}\int_0^{t_2} f(t_1,\ldots,t_n)\mathrm{d}W(t_1)\mathrm{d}W(t_2)\ldots\mathrm{d}W(t_{i-1}),\ \ t_i \in \big[0, t_{i+1}\big],$$

is $\mathbb{F}$-adapted and square integrable with respect to $\mathrm{d}P \times \mathrm{d}t_i$, $J_n(f)$ is well defined. This means that $J_n \in L^2(P)$, space of square integrable random variables.

If we take t to be time in applications, then S_n denotes monotonic increase in time, and the time interval $[0, T]$ can be divided into equidistant time intervals. If g and h deterministic functions defined on S_m and S_n, respectively, with $g \in L^2(S_m)$ and $h \in L^2(S_n)$, , then it can be shown that Di Nunno et al. (2019) for $m < n$,

$$E\big[J_m(g)J_n(h)\big] = 0, \tag{2.3}$$

and if both g and $h \in L^2(S_n)$, i.e., $m = n$,

$$E\big[J_m(g)J_n(h)\big] = (g, h)_{L^2(S_n)}, \tag{2.4}$$

where,

$$(g, h)_{L^2(S_n)} := \int gh\mathrm{d}t_1\ldots\mathrm{d}t_n, \tag{2.5}$$

the inner product of $L^2(S_n)$. Further,

$$\|J_n(h)\|_{L^2(P)} = \|h\|_{L^2(P)}. \tag{2.6}$$

It should also be noted that J_n is a linear operator.

Another definition of the n-fold Itô iterated integral is defined on the symmetric functions. There are very good mathematical reasons why symmetric functions are needed but we do not discuss the topic in this book, and it is sufficient to say that there is excellent literature of symmetrization of functions (Egge, 2019). If $g \in L^2(S_m)$ belongs to the symmetric function space $\tilde{L}^2([0, T]^n)$, the n-fold iterated integral, $I_n(g)$, , is defined as (Di Nunno et al., 2019),

$$I_n(g) := \int g(t_1,\ldots,t_n)\mathrm{d}W(t_1)\mathrm{d}W(t_2)\ldots\mathrm{d}W(t_{n-1})\mathrm{d}W(t_n), \tag{2.7}$$

integrating over $[0, T]^n$.

An important result is given by,

$$I_n(g) = n!J_n(g). \tag{2.8}$$

For the same domain of t, $I_n(g)$ is a much larger quantity than $J_n(g)$ and we can use Eq. (2.8) effectively in computations. It can also be shown that,

$$\|I_n(g)\|^2_{L^2(P)} = n!\|g\|^2_{L^2([0,T]^n)} \tag{2.9}$$

for all $g \in \tilde{L}^2([0, T]^n)$.

Further, if $g \in \tilde{L}^2([0, T]^n)$, and $h \in \tilde{L}^2([0, T]^n)$, then

$$E\big[I_n(g)I_n(h)\big] = \begin{cases} 0, & n \neq m \\ (g, h)_{\tilde{L}^2([0,T]^n)}, & n = m \end{cases} \quad (m, n = 1, 2, \ldots) \tag{2.10}$$

where

$$(g, h)_{\tilde{L}^2([0,T]^n)} = n!(g, h)_{L^2(S_n)}. \tag{2.11}$$

The Itô iterated integral I_n defined above is the fundamental object in defining further mathematical structures that may be used in applications. With the appropriate "test functions" from the theory of distributions, I_n can be computed for $L^2([0, T])$. A test function is a rapidly decreasing function with bounds; it can be shown that Lebesgue integral of nondifferentiable continuous functions can be integrated by using test functions, and this important result is used in the theory of partial differential equations using special spaces that will be given later in the chapter. Readers are referred to Debnath and Mikusinski (2005) for further details. One such useful test functions to evaluate the I_n are Hermite functions. Hermite polynomials are defined by,

$$h_n(x) = (-1)^n e^{\left(\frac{1}{2}x^2\right)} \frac{d^n}{dx^n}\left(e^{-\left(\frac{1}{2}x^2\right)}\right), \quad x \in \mathbb{R}, \ n = 0, 1, 2, \ldots. \tag{2.12}$$

In physics, Hermite polynomials are defined slightly differently but have the same characteristics. We use the definition given in Eq. (2.12) for the computations in this book. Hermite polynomials and Gaussian density functions are related, forming an orthogonal basis for $L^2(\mathbb{R}, \mu(dx))$ with a measure given by, (See Di Nunno et al. (2019).)

$$\mu(dx) = \frac{1}{2\pi} e^{\left(\frac{1}{2}x^2\right)dx}. \tag{2.13}$$

The Hermite functions $hf_k(x)$ are defined by,

$$hf_k(x) = \pi^{-(1/4)}((k-1)!)^{-1/2}\left(e^{-\left(\frac{1}{2}x^2\right)}\right)h_{k-1}\left(\sqrt{2}x\right), \quad k = 1, 2, \ldots. \tag{2.14}$$

These are the main properties of the Hermite functions:

- They belong to the Schwartz space of rapidly decreasing smooth (C^∞) real valued functions on $\mathbb{R}^d$ which is used to construct the white noise probability space, and

- The set $\{hf_k(x)\}_{k=1}^{\infty}$ forms an orthonormal basis for $L^2(\mathbb{R})$, $x \in \mathbb{R}$, and $hf_k \in S(\mathbb{R})$ for all k.

Note that the Hermite functions can be defined on $\mathbb{R}^d$, and d is called the parameter dimension that could be space or time–space or time dimensions of the system we model. In Sect. 2.5 of this chapter, we compute a few Hermite polynomials and functions, and we can see how rapidly Hermite functions decay visually. We can define the action ω on hf_k as follows:

$$\theta_k(\omega) := \langle \omega, hf_k \rangle = w_{hf_k}(\omega) = \int_{-l}^{l} hf_k(x) \mathrm{d}W(x), \quad \omega \in \Omega. \tag{2.15}$$

Here $[-l \le x \le l] \in \mathbb{R}$. Equation (2.15) is an Itô integration of a deterministic function which gives a single value for each realization. For a large number of realizations of $\theta_k(\omega)$, smooth white noise mediated by $hf_k(x)$ can be obtained. We discuss these computed realizations in Sect. 2.5 and Mathematica® (Wolfram, Version 14) code is given in Appendix 2.1.

If $\mathcal{J}$ denote the set of all multi-indices $\alpha = (\alpha_1, \alpha_2, \ldots, \alpha_m)$, $m = 1, 2, \ldots$, of nonnegative integers α_i. (See Appendix 2.2 for a summary of index notation.) We can define a stochastic function of ω for

$$\alpha = (\alpha_1, \alpha_2, \ldots, \alpha_m) \in \mathcal{J}, \quad \alpha \neq 0,$$

$$H_\alpha(\omega) := \prod_{j=1}^{m} h_{\alpha_j}\left(\theta_j(\omega)\right), \quad \omega \in \Omega. \tag{2.16}$$

Equation (2.16) is an important stochastic function, and for example, for $\alpha = (3, 0, 2)$, following (Di Nunno et al., 2019),

$$\begin{aligned} H_{(3,0,2)}(\omega) &= h_3(\theta_1(\omega)) h_0(\theta_2(\omega)) h_2(\theta_3(\omega)) \\ &= \left(\theta_1^3 - 3\theta_1\right)\left(\theta_3^2 - 1\right). \end{aligned} \tag{2.17}$$

On the other hand,

$$H_{(2,0,3)}(\omega) = h_2(\theta_1(\omega)) h_0(\theta_2(\omega)) h_3(\theta_3(\omega)) = \left(\theta_1^2 - 1\right)\left(\theta_3^3 - 3\theta_3\right),$$

which produce different realizations. We compute $H_{(3,0,2)}$ and $H_{(2,0,3)}(\omega)$ in Sect. 3. 5 to illustrate realizations graphically.

Extending the previous arguments, it can be shown that if $\xi_1, \xi_2, \ldots, \xi_m$ are orthonormal functions in $L^2([0, T])$, we can express the n-fold iterated integral I_n as follows (Di Nunno et al., 2019; Ito, 1951):

$$I_n\left(\xi_1^{\otimes\alpha_1}\hat{\otimes}\cdots\hat{\otimes}\xi_m^{\otimes\alpha_m}\right)=\prod_{k=1}^{m}h_{\alpha_k}\left(\int_0^T\xi_k(t)\mathrm{d}W(t)\right),\tag{2.18}$$

with index notation and

$$\alpha_1+\alpha_2+\alpha_3\cdots=n,\quad \alpha_k\in\{0,1,2,\ldots\}\quad \text{for all}\,k.$$

We can simplify this equation for the deterministic functions for potential applications.

The tensor product $g_1\otimes g_2$ of two deterministic functions is defined by

$$(g_1\otimes g_2)(x_1,x_2)=g_1(x_1)g_2(x_2)$$

and the symmetrized tensor product, $g_1\hat{\otimes}g_2$ is the symmetrization of $g_1\otimes g_2$.

One of the most useful results for us in our development is as follows:

$$n!\int_0^T\int_0^{t_n}\cdots\int_0^{t_3}\int_0^{t_2}g(t_1,\ldots,t_n)\mathrm{d}W(t_1)\mathrm{d}W(t_2)\ldots\mathrm{d}W(t_{n-1})\mathrm{d}W(t_n)=\|g\|^n h_n\left(\frac{\theta}{\|g\|}\right),\tag{2.19}$$

for the tensor power of $g\in L^2([0,T])$; where $\|g\|=\|g\|_{L^2([0,T])}$ and,

$$\theta=\int_0^T g(t)\mathrm{d}W(t).\tag{2.20}$$

In Eq. (2.19), we express an iterated Itô integral of $g(t)$ at distinct times ($t_1\le t_2\le\ldots\le t_n$) in terms of the nth Hermite polynomial and the norm of g in $L^2([0,T])$; θ is an Itô integral over the entire interval ($[0,T]$). This formula is useful in computations.

Now we come to a significant result: the Wiener–Itô Chaos expansion of any $\mathcal{F}_T$ measurable random variable in $L^2(P)$, ξ, which can be expressed as a summation of the n-fold iterated Itô integral. For any such ξ, there exists a unique sequence of deterministic, symmetric functions, $\{f_n\}_{n=0}^{\infty}$, $f_n\in\tilde{L}^2([0,T]^n)$ such that,

$$\xi=\sum_{n=0}^{\infty}I_n(f_n)\tag{2.21}$$

with the norm in $L^2(P)$,

$$\|\xi\|^2_{L^2(P)}=\sum_{n=0}^{\infty}n!\|f_n\|^2_{\tilde{L}^2([0,T]^n)}.\tag{2.22}$$

Note that f_ns are symmetric functions belonging to $\tilde{L}^2([0, T]^n)$. The Wiener–Itô Chaos expansions are central to stochastic analysis as it provides a way to express stochastic variables in terms of computable Itô integrals.

2.3 The Skorohod Integral and the Malliavin Derivative

2.3.1 White Noise Probability Space

Following Hida's development of white noise probability space (Hida, 1980), let $S = S(\mathbb{R}^d)$ be the Schwartz space of rapidly decreasing smooth $C^\infty(\mathbb{R}^d)$ real functions; the S is a Frechet space with respect to the family of seminorms (Holden et al., 2010):

$$\|f\|_{K,\alpha} := \sup_{x \in \mathbb{R}^d} \left\{ \left(1 + |x|^K\right) |\partial^\alpha f(x)| \right\}$$

where $K = 0, 1, \ldots,$ and $\alpha = (\alpha_1, \ldots, \alpha_d)$ is a multi-index with $\alpha_j = 0, 1, \ldots$ $(j = 1, \ldots, d)$ and

$$\partial^\alpha f := \frac{\partial^{|\alpha|}}{\partial x_1^{\alpha_1} \ldots \partial x_d^{\alpha_d}} f \quad \text{for } |\alpha| = \alpha_1 \ldots \alpha_d.$$

If we define the dual space of S as $S' = S'(\mathbb{R}^d)$, which is the space of tempered distributions, we can define the "action" of $\omega(\in S')$ on $\phi(\in S)$ as $\langle \omega, \phi \rangle$ or $\omega(\phi)$.

Given a measure m on $\mathbb{R}^d$ then,

$$\langle \omega, \phi \rangle = \int \phi(x) \mathrm{d}m(x)$$

over $\mathbb{R}^d$ and, in particular, if the measure m is concentrated on $x_0 \in \mathbb{R}^d$, then

$$\langle \omega, \phi \rangle = \phi(x_0)$$

is the evaluation of ϕ at $x_0 \in \mathbb{R}^d$. If $\mathcal{B}$ is the Borel subsets of $S'(\mathbb{R}^d)$, then by the Bochner–Minlos–Sazonov theorem (Holden et al., 2010), there exists probability measure P on Ω such that

$$\int \mathrm{e}^{-i\langle \omega, \phi \rangle} P(\mathrm{d}\omega) = \mathrm{e}^{-\frac{1}{2}\|\phi\|^2},$$

integrating over Ω, $\phi \in S$, where

$$\|\phi\|^2 = \|\phi\|^2_{L^2(\mathbb{R})} = \int |\phi(x)|^2 \mathrm{d}x.$$

P is called the white noise probability measure and $(\Omega, \mathcal{F}, P) = \big(S', \mathcal{B}, P\big)$ is defined as the white noise probability space (Di Nunno et al., 2019; Holden et al., 2010). This is the space we use to develop the stochastic models in this monograph. We can construct a smooth white noise process $\big(w : S \times S' \to \mathbb{R}\big)$ by using $\langle \omega, \phi \rangle$, i.e., the action previously defined can be considered as a smoothed white noise process in the white noise probability space. It can be shown that smoothed white noise process, $w_\phi(\omega)$, can be computed by using (Di Nunno et al., 2019),

$$w_\phi(\omega) = \int \phi(t)\mathrm{d}W(t, \omega), \quad \omega \in \Omega, \quad \phi \in L^2(\mathbb{R}),$$

which is an Itô-integral.

There is a very useful connection between the action $\langle \omega, \phi \rangle$ and the n-fold iterated Itô integral if ϕ is defined as the Hermite function.

Let e_k be the kth Hermite function defined as given before, (See Eq. (2.14).)

$$e_k(x) = \pi^{-(1/4)}((k-1)!)^{-1/2}\Big(\mathrm{e}^{-(\frac{1}{2}x^2)}\Big)h_{k-1}\Big(\sqrt{2}x\Big), \quad k = 1, 2, \ldots, \text{ and}$$

$\{e_k\}_{k\geq 1}$ is an orthonormal basis for $L^2(\mathbb{R})$ and $e_k \in S(\mathbb{R})$ for all k (Holden et al., 2010).

We can define, an action between ω and e_k,

$$\theta_k(\omega) := \langle \omega, e_k \rangle = \int e_k(x)\mathrm{d}W(x, \omega), \quad x \in \mathbb{R}, \quad \omega \in \Omega. \tag{2.23}$$

We apply Eq. (2.23) in our computations in later chapters, and $\theta_k(\omega)$ is an important mathematical object in defining the Wick products.

2.3.2 *Skorohod Integral*

From practical applications perspective, the Itô integral may have limitations in predictive modeling. As the Itô integral is nonanticipating, i.e., $\mathbb{F}$ adapted, and involves the Itô calculus, the Skorohod integral, which is applied to anticipating processes and also has the zero-expectation property, would be a better choice in the modeling of physical systems. Mainly using the Wick product, we could develop models using the white noise probability space (Di Nunno et al., 2019; Holden et al., 2010). These models do not require the Itô calculus as we see in our applications. Further this integral is an adjoint operator to the Malliavin derivative we introduce next (Di Nunno et al., 2019).

A simple way to think about the Skorohod integral is as a summation of the n-fold iterative Itô integrals for symmetric functions. Resorting to the previous definition of Wiener–Itô Chaos expansion of random variable $\xi \big(= \sum_{n=0}^{\infty} I_n(f_n)\big)$ with the given norm, based on symmetric functions $\{f_n\}$, the Skorohod integral of ξ can be defined as,

$$\delta(\xi) := \int_0^T \xi\, \delta W(t) := \sum_0^\infty I_{n+1}(f_n),$$

where f_ns are the symmetric functions at $t \in [0, T]$. We write $\xi \in \mathrm{Dom}(\delta)$ if the above summation converges in $L^2(P)$. (See Di Nunno et al. (2019) for examples of derivation of the Skorohod integral.) $\delta(\xi)$ is a linear operator which act as an integral with all the properties attributed to an integral. When ξ is $\mathbb{F}$ adapted, i.e., nonanticipating, the Skorohod integral and the Itô integral are the same. In some modeling situations when the random variables are nonanticipating, we could simplify the calculations.

2.3.3 *Malliavin Derivative*

By using the white noise probability space discussed above, the Malliavin derivative $D_t F$ of a random variable $F : S' \to \mathbb{R}$ is a stochastic gradient of F at t. But the Malliavin derivative can also be written in terms of the n-fold Itô integral I_n:

$$D_t F = \sum_{n=1}^\infty n I_{n-1}(f_n(.,t)). \tag{2.24}$$

Intuitively we see the link between the Malliavin derivative and the Skorohod integral defined previously. The problem is however to find a suitable symmetric function f which is suitable for the process we want to model. This is where the anticipating Wick calculus paves the way.

It should be noted that our overly simplified description of the Malliavin derivative is not mathematically rigorous. However, one can think of this derivative as a model of the rate of change of F in $\omega(\in S')$ direction. However, the Malliavin derivative is valid, strictly speaking, if $F \in \mathbb{D}_{1,2}$, the norm of F follows the relationship:

$$\|F\|^2_{\mathbb{D}_{1,2}} := \sum_{n=1}^\infty n n! \|f_n\|^2_{L^2([0,T]^n)} < \infty. \tag{2.25}$$

Once this condition is satisfied, which we assume that it does when the functions are used appropriately, the Malliavin derivative follows similar rules like in normal calculus. We make use of these rules as we develop our computational models. Another important theorem we make use of is the Clark–Ocone formula, which expresses F in terms of an Itô integrals. There is a nice discussion on the difference between Clark–Ocone formula and the Itô representation theorem in Chap. 4 of Di Nunno et al. (2019). As can be expected, there are strong relationships between the Malliavin derivative and the Skorohod integral, the aspects of which are highly technical; but we state these relationships whenever we use them in subsequent chapters. (See Di Nunno et al. (2019) for further details.)

2.4 The Wick Product

2.4.1 Useful Mathematical Objects

If $\mathcal{J}$ is the set of all finite multi-indices $\alpha = (\alpha_1, \alpha_2, \ldots, \alpha_m)$, $m = 1, 2, \ldots$, of nonnegative integers the set of all multi-indices α_i, we define earlier,

$$H_\alpha(\omega) := \prod_{j=1}^{m} \left(\theta_j(\omega)\right), \quad \omega \in \Omega. \tag{2.26}$$

Following Itô (Di Nunno et al., 2019; Hida, 1980; Holden et al., 2010; Ito, 1951), the iterated Itô integral can be expressed as,

$$I_m\left(e^{\hat{\otimes}\alpha}\right) = \prod_{j=1}^{m} h_{\alpha_j}\left(\theta_j(\omega)\right) = H_\alpha(\omega) \tag{2.27}$$

where h_{α_j} is the α_jth Hermite polynomial as defined previously. $\otimes$ and $\hat{\otimes}$ denote the tensor product and the symmetrized tensor product, respectively. One simple result that is useful to us later, is, if f and g are real functions on $\mathbb{R}^1$,

$$(f \otimes g)(x_1, x_2) = f(x_1)g(x_2), \text{ and} \tag{2.28}$$

$$\left(f \hat{\otimes} g\right)(x_1, x_2) = \frac{1}{2}\left[f(x_1)g(x_2) + f(x_2)g(x_1)\right], \tag{2.29}$$

this is because the symmetrization of $f(x_1)g(x_2)$ is

$$\frac{1}{2}\left[f(x_1)g(x_2) + f(x_2)g(x_1)\right].$$

According to Theorem 5.2 of Di Nunno et al. (2019), the Wiener–Itô Chaos expansion can be expressed in terms of $H_\alpha(\omega)$, which is a family of functions which forms an orthogonal basis of $L^2(P)$. The Wiener–Itô expansion is expressed, for all $\mathcal{F}_T$-measurable $X \in L^2(P)$, as

$$X = \sum_{\alpha \in \mathcal{J}} c_\alpha H_\alpha \tag{2.30}$$

and

$$\|X\|_{L^2(P)}^2 = \sum_{\alpha \in \mathcal{J}} \alpha! c_\alpha^2, \tag{2.31}$$

where $\alpha! = \alpha_1!\alpha_2!\ldots\alpha_m!$ for $\alpha = (\alpha_1, \alpha_2, \ldots, \alpha_m)$.

Like the test function space of $S(\mathbb{R}^d)$ and its dual $S'(\mathbb{R}^d)$, we have families of spaces related to stochastic test functions and distributions. One such set of spaces is called Kondratiev spaces, and another family is called Hida spaces. The technical aspects of these spaces are given by Holden et al. (2010) and we do not describe them in detail because we can use Hida test function space (S) and the Hida distribution space (S^*). According to Proposition 2.3.7 of Holden et al. (2010), for arbitrary dimension m and the measure μ_m: The Hida test function space $(S)^N$ consists of those functions

$$f = \sum_{\alpha \in \mathcal{J}} c_\alpha H_\alpha \in L^2(\mu_m) \tag{2.32}$$

with $c_\alpha \in \mathbb{R}^N$ such that

$$\sup_\alpha \left\{c_\alpha^2 \alpha! (2N)^{k\alpha}\right\} < \infty \text{ for all } k < \infty.$$

The Hida distribution space $(S^*)^N$ of all formal expansions, $F = \sum_{\alpha \in \mathcal{J}} b_\alpha H_\alpha$ with $b_\alpha \in \mathbb{R}^N$ such that $\sup_\alpha \left\{b_\alpha^2 \alpha! \, (2N)^{-q\alpha}\right\} < \infty$ for some $q < \infty$. If $N = 1$ (1 parameter dimension), we write $(S)^1 = (S)$ and $(S^*)^N = (S^*)$. (S^*) contains singular or pointwise white noise of dimension. (See details in Holden et al. (2010).)

2.4.2 Introduction to the Wick Product

The Wick product is an important tool in stochastic analysis. If two variables are stochastic and can be considered as elements of (S^*), then their multiplication is given by the Wick product. We define the Wick product as follows:

$$\text{If } X = \sum_{\alpha \in \mathcal{J}} a_\alpha H_\alpha \in (S^*), \; Y = \sum_{\beta \in \mathcal{J}} b_\beta H_\alpha \in (S^*),$$

then the Wick product $X \diamond Y$ of X and Y is defined as,

$$X \diamond Y := \sum_{\alpha,\beta} a_\alpha b_\beta H_{\alpha+\beta} = \sum_\gamma \left(\sum_{\alpha+\beta=\gamma} a_\alpha b_\beta \right) H_\gamma. \tag{2.33}$$

(See Di Nunno et al. (2019) for details.)

One can show that $X, Y \in (S)^* \Rightarrow X \diamond Y \in (S)^*$ and if $X, Y \in (S) \Rightarrow X \diamond Y \in (S)$. However, if $X, Y \in L^2(P)$, that does not lead to $X \diamond Y \in L^2(P)$ in general. Following (Di Nunno et al., 2019), some of the useful properties of the Wick product are listed below:

- If at least one of the variables X and Y are deterministic, then $X \diamond Y = X \cdot Y$. The Wick product and the ordinary product (ω-pointwise) are the same. This property is used in our calculations to simplify the equations.
- The singular white noise w can be defined using the Wick product.
- There is a fundamental relation between Itô/Skorohod integrals and the Wick products:

$$\int Y(t)\delta W(t) = \int Y(t) \diamond W(t)dt \tag{2.34}$$

integrating in $\mathbb{R}$.

Equation (2.34) is a useful result in stochastic differential equations (SDEs) as we see later in the applications. The strong solutions of SDEs can be obtained by using the Wick product for many situations.

Based on this rudimentary exposition in the chapter, we build the machinery of mathematics for the Wick algebra in the next chapter.

2.5 Computations

In this section, we visualize some of the mathematics discussed earlier to obtain some insights into the behavior of deterministic and stochastic functions. Figure 2.1 shows the Hermite polynomials $h_n(x)$ for $n = 2, 4$, and 6 for the range of $x = [-3, 3]$, and when we enlarge the range of x to $[-10, 10]$, the relative magnitude of $h_4(x)$ is significantly higher than $h_2(x)$ beyond $x > 5$ and $x < -5$ (Fig. 2.2).

But the behaviors of Hermite functions are quite different to those of polynomials (Fig. 2.3) and note the rapidly decreasing nature of them when the ranges are large enough (Fig. 2.4).

When $n = 5$ and 6, it is interesting to note that the rapidly decreasing nature is confined to very similar ranges as those for $n = 1, 2, 3$, and 4 (Fig. 2.5).

We compute $\theta_k(\omega)$, which is the smooth white noise or the action of ω on hf_k, for $k = (1, 2, 3, 4, 5, 6)$ next. Figure 2.6 depicts 100 $\theta_1(\omega)$ realizations for the range of $x = [0, 1]$ with $\Delta x = 0.0001$.

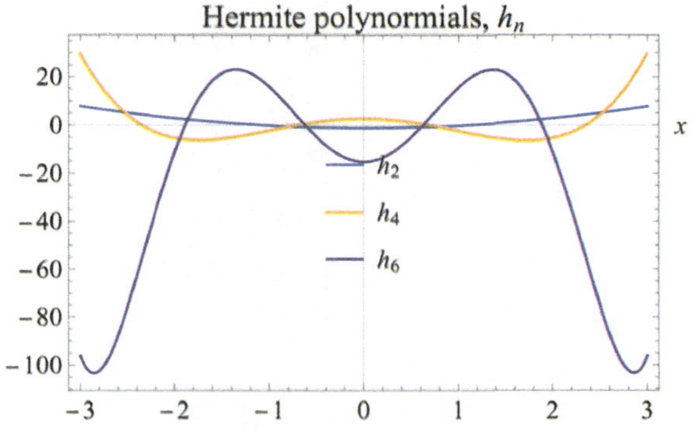

Fig. 2.1 Hermite polynomials when $n = 2, 4, 6$

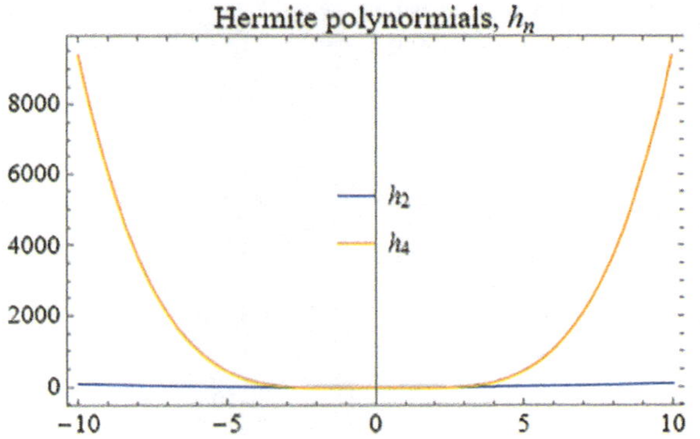

Fig. 2.2 Hermite polynomials for a larger range

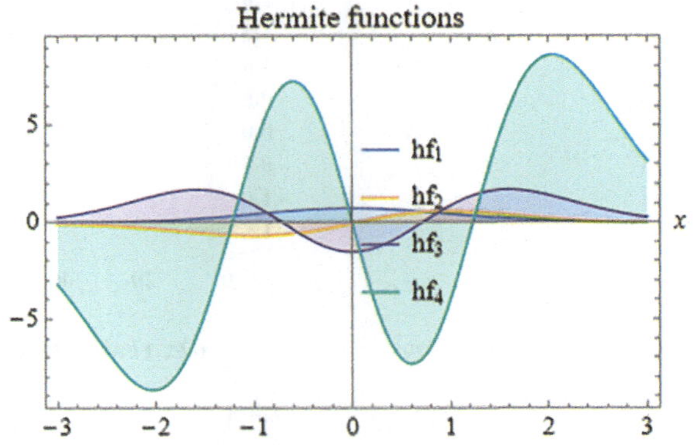

Fig. 2.3 Hermite functions for $n = 1, 2, 3, 4$

Figure 2.7 shows $\theta_2(\omega)$ for the same parameters as before and note that there is hardly any changes of amplitudes between Figs. 2.6 and 2.7.

Similarly, Fig. 2.8 shows all $\theta_1(\omega)$, $\theta_2(\omega)$, and $\theta_3(\omega)$ overlayed on each other.

Fig. 2.4 Hermite functions for a larger range of x

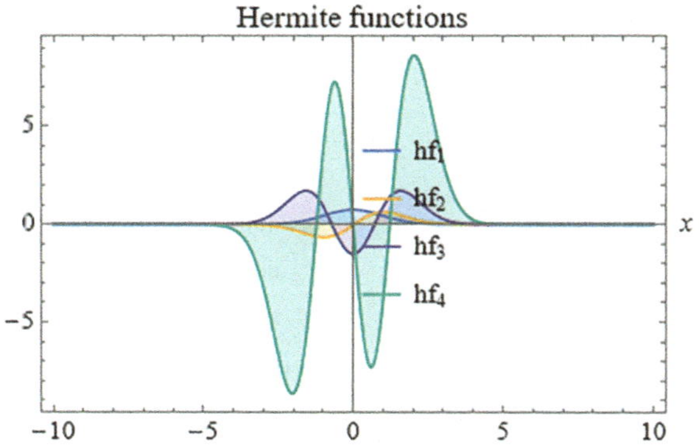

Fig. 2.5 Rapidly decreasing nature of Hermite functions for larger n

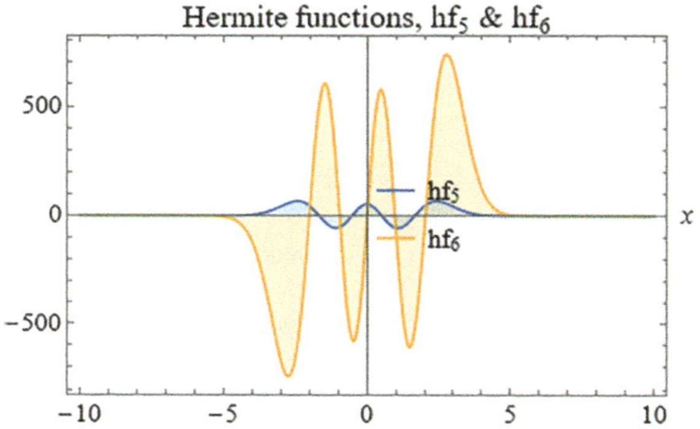

Fig. 2.6 $\theta_1(\omega)$ realizations for $x = [0, 1]$

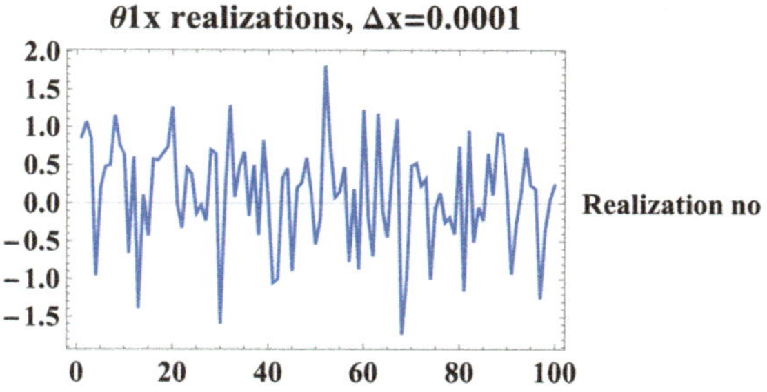

Fig. 2.7 $\theta_2(\omega)$ realizations for $x = [0, 1]$

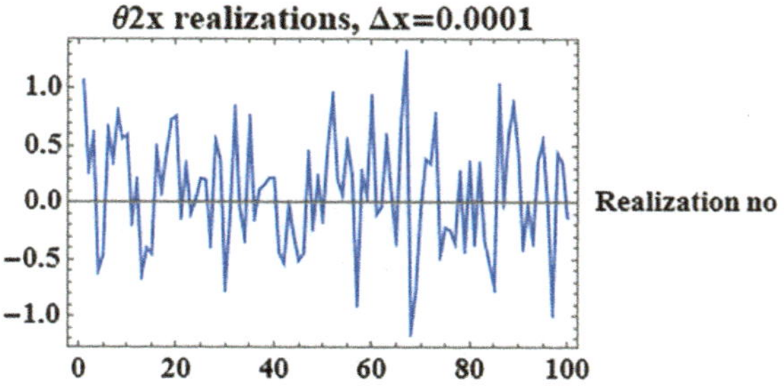

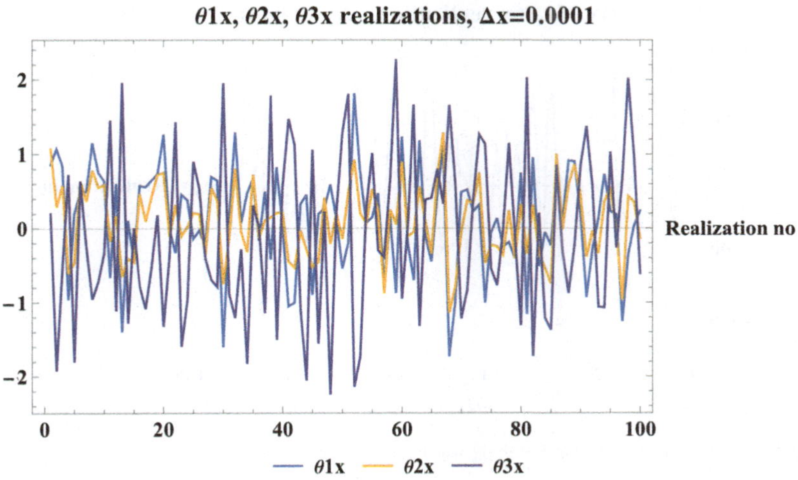

Fig. 2.8 $\theta_1(\omega)$, $\theta_2(\omega)$, and $\theta_3(\omega)$ together

Fig. 2.9 $\theta_4(\omega)$ realizations for $x = [0, 1]$

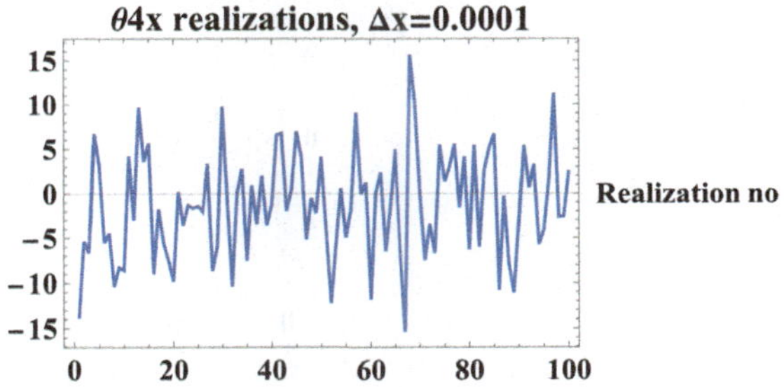

The peaks and troughs of $\theta_3(\omega)$ are higher and lower, respectively, than those of the other two graphs, but generally shows white noise behavior overall. However, peaks and troughs of $\theta_4(\omega)$ are significantly higher and lower, respectively, Fig. 2.9 to those of in Fig. 2.8 and this trend continues for $\theta_5(\omega)$, and $\theta_6(\omega)$ (Fig. 2.10).

The increase in amplitudes is expected given the graph of hf_4 in Fig. 2.5. Then there is no surprise to see large amplitudes in $\theta_5(\omega)$ and $\theta_6(\omega)$ as shown in Fig. 2.10.

Now to answer the question of how these actions of ω on hf_k affect the stochastic function $H_\alpha(\omega)$, we compute $H_{(3,0,2)}(\omega)$ and $H_{(2,0,3)}(\omega)$, and Fig. 2.11 shows 100 realizations overlaid on each other.

No significant differences in amplitudes can be seen of these functions, but we can expect multi-index α to have an effect in different combinations. These amplitudes drive the coefficients in the Wiener–Itô Chaos expansion.

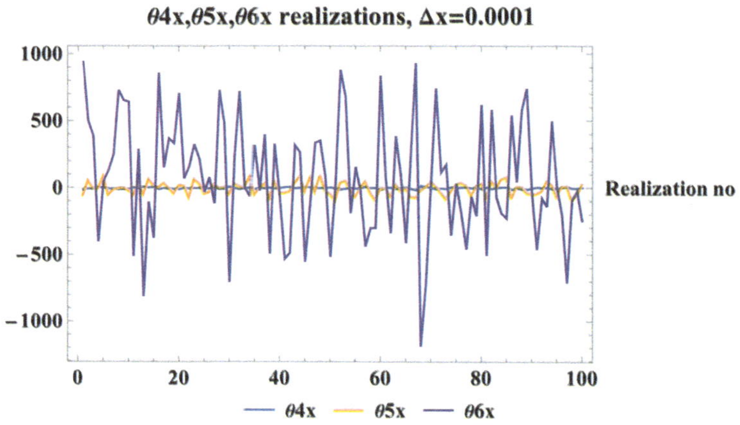

Fig. 2.10 $\theta_5(\omega)$ and $\theta_6(\omega)$ along with $\theta_4(\omega)$, blue graph along the x-axis

Fig. 2.11 100 realizations of $H_{(3,0,2)}(\omega)$ and $H_{(2,0,3)}(\omega)$

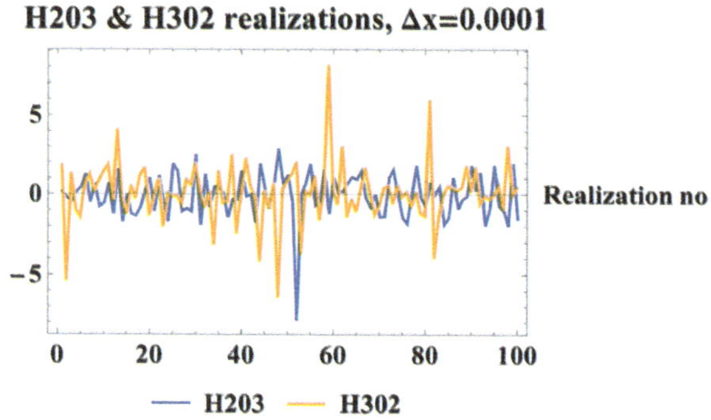

Appendix 2.1: Mathematica® Code Without Outputs

(*We compute Hermite polynomials and Hermite functions first to show the behavior of them visually. Note that the Hermite polynomials mathematicians use are slightly different from the ones physicists use. Mathematica uses the form physicists use. Therefore, we use the definitions given in Chap. 2.*)

```
HermitePoly[n_, x_]
:= Expand[((− 1)^n)Exp[(1/2)x^2]D[Exp[(− 1/2)x^2], {x, n}]];

Plot[Evaluate[{HermitePoly[2, x], HermitePoly[4, x],
    HermitePoly[6, x]}], {x, − 3, 3},

  Frame → True, Axes → True, PlotRange → Full,
```

```
PlotLabel → "Hermite polynormials, h_n",

AxesLabel → {HoldForm[x]}, ColorFunction → Automatic,

LabelStyle → {FontFamily
        → Times New Roman, 14, GrayLevel[0]},

PlotLegends → Placed[{h2, h4, h6}, {Center, Center}]]

(*Plotting for a larger range*)

Plot[Evaluate[{HermitePoly[2, x], HermitePoly[4, x]}], {x, - 10, 10},

Frame → True, Axes → True,

PlotRange → Full, PlotLabel → "Hermite polynormials, h_n",

AxesLabel → {HoldForm[x]}, ColorFunction → Automatic,

LabelStyle → {FontFamily
        → Times New Roman, 14, GrayLevel[0]},

PlotLegends → Placed[{h2, h4}, {Center, Center}]]

(*For Hermite functions as given in Chap. 2, make a substitution, y = Sqrt[2]x.*)

HermiteFunc[n_, y_]

:= Expand[π^(-1/4)((n - 1)!)Exp[-y^2/4]

HermitePoly[(n - 1), y]];

Plot[Evaluate[{HermiteFunc[2, y], HermiteFunc[3, y], HermiteFunc[4, y]}],
    {y, - 3, 3},

PlotRange → Full, Frame → True, Axes → True,
```

```
PlotRange → Full,

PlotLabel → "Hermite functions, hf''_n,

AxesLabel → {HoldForm[y]}, ColorFunction → Automatic,

LabelStyle → {FontFamily
→ "Times New Roman", 14, GrayLevel[0]}, PlotLegends
→ Placed[{hf2, hf3, hf4}, {Center, Center}]]
```

(*Now we evaluate Hermite functions as a function of x*)

```
hf1 = HermiteFunc[1, y]; hf2 = HermiteFunc[2, y];

hf3 = HermiteFunc[3, y]; hf4 = HermiteFunc[4, y];

hf5 = HermiteFunc[5, y]; hf6 = HermiteFunc[6, y];

y = √2x;

Plot[{hf1, hf2, hf3, hf4}, {x, -3, 3}, Filling → Axis, PlotRange
    → Full,

PlotRange → Full, Frame → True, Axes → True,

PlotLabel → "Hermite functions",

AxesLabel → {HoldForm[x]},

ColorFunction → Automatic,

LabelStyle → {FontFamily
→ "Times New Roman", 14, GrayLevel[0]}, PlotLegends
→ Placed[{hf1, hf2, hf3, hf4}, {Center, Right}]]
```

(*Plotting for a larger range*)

```
Plot[{hf1, hf2, hf3, hf4}, {x, – 10, 10}, Filling → Axis,

PlotRange → Full, PlotRange → Full, Frame → True,

Axes → True, PlotLabel → "Hermite functions",

AxesLabel → {HoldForm[x]}, ColorFunction → Automatic,

LabelStyle → {FontFamily
→ "Times New Roman", 14, GrayLevel[0]}, PlotLegends
→ Placed[{hf1, hf2, hf3, hf4}, {Center, Right}]]

Plot[{hf5, hf6}, {x, – 3, 3}, Filling → Axis, PlotRange → Full,

Frame → True, Axes → True,

PlotLabel → "Hermite functions, hf5 and hf6",

AxesLabel → {HoldForm[x]}, ColorFunction → Automatic,

LabelStyle → {FontFamily
→ "Times New Roman", 14, GrayLevel[0]}, PlotLegends
→ Placed[{hf5, hf6}, {Center, Right}]]

(*Plotting for a larger range*)

Plot[{hf5, hf6}, {x, – 10, 10}, Filling → Axis, PlotRange → Full,

Frame → True, Axes → True,

PlotLabel → "Hermite functions, hf5 and hf6",

AxesLabel → {HoldForm[x]}, ColorFunction → Automatic,

LabelStyle → {FontFamily
→ "Times New Roman", 14, GrayLevel[0]}, PlotLegends
```

$\rightarrow$ Placed[{hf$_5$, hf$_6$}, {Center, Right}]]

(*The following part could be run independently to the previous part as a separate notebook*)

(*$n = 0, 1, 2 \ldots; k = 1, 2, \ldots$*)

(*Hermite functions form an orthonormal basis for $L^2(R)$ and HermiteFunc[k, x] is in $S(R)$. θ_k is the action defined by $\int \theta_k(x)\mathrm{d}W$, hence we compute $\theta_k(x)$. To start we compute Hermite polynomials and Hermite functions for y and substitute $y =$ Sqrt(2) x to get the functions in x*)

$$\text{HermiteFunc}[n\_, y\_] := \text{Expand}[\pi^{(-1/4)}((n-1)!)\text{Exp}[-y^2/4]$$

$$\text{HermitePoly}[(n-1), y]];$$

$$\text{hf1} = \text{HermiteFunc}[1, y];$$

$$\text{hf2} = \text{HermiteFunc}[2, y];$$

$$\text{hf3} = \text{HermiteFunc}[3, y];$$

$$\text{hf4} = \text{HermiteFunc}[4, y];$$

$$\text{hf5} = \text{HermiteFunc}[5, y];$$

$$\text{hf6} = \text{HermiteFunc}[6, y];$$

(*Then we compute the Wiener increments along x-axis*)

$$y = \sqrt{2}x;$$

$$\Delta x = 0.0001;\ \text{N} = 10,000;$$

(*Compute 100 realizations of Wiener increments*)

$$\Delta\text{WListMatrix}$$
$$= \text{Table}[N[\text{Random}[\text{NormalDistribution}[0., \text{Sqrt}[\Delta n]]], 32], \{\text{N}\}, \{100\}];$$

(*Compute $\theta_1 x$ realizations for Hermite function hf1; Each realization produces a single value over $x = [0, 1]$*)

$$\theta 1\text{x} = \text{Table}[\text{Sum}[x = (i-1)\Delta x;$$

$$\text{Evaluate}\left[\text{hf1}\,\Delta\text{WListMatrix}\left[\left[i,j\right]\right]\right], \{i, 1, \text{N}\}\right], \{j, 1, 100\}\right];$$

$$\text{ListLinePlot}[\theta 1\text{x}, \text{Frame} \to \text{True}, \text{Axes} \to \text{True}, \text{Joined} \to \text{True},$$

$$\text{PlotLabel} \to {''}\theta\, 1x\,\text{realizations}, \Delta x = 0.0001{''},$$

$$\text{AxesLabel} \to \{\text{HoldForm}[\text{Realization no}]\},$$

$$\text{ColorFunction} \to \text{Automatic},$$

$$\text{LabelStyle} \to \{\text{FontFamily} \to {''}\text{Times New Roman}{''}, 14, \text{GrayLevel}[0], \text{FontWeight} \to \text{Bold}\}\Big]$$

(*Compute $\theta_2 x$ realizations for Hermite function hf2; Each realization produces a single value over $x = [0, 1]$*)

$$\theta 2x = \text{Table}[\text{Sum}[x(i-1)\Delta x\,;$$

$$\text{Evaluate}\left[\text{hf2}\,\Delta\text{WListMatrix}\left[\left[i,j\right]\right]\right], \{i, 1, \text{N}\}\right], \{j, 1, 100\}\right];$$

$$\text{ListLinePlot}[\theta 2x, \text{Frame} \to \text{True}, \text{Axes} \to \text{True}, \text{Joined} \to \text{True},$$

$$\text{PlotLabel} \to {''}\theta 2x\,\text{realizations}, \Delta x = 0.0001{''},$$

$$\text{AxesLabel} \to \{\text{HoldForm}[\text{Realization no}]\},$$

$$\text{ColorFunction} \to \text{Automatic},$$

$$\text{LabelStyle} \to \{\text{FontFamily} \to {''}\text{Times New Roman}{''}, 14, \text{GrayLevel}[0], \text{FontWeight} \to \text{Bold}\}\Big]$$

(*Compute $\theta_3 x$ realizations for Hermite function hf3; Each realization produces a single value over $x = [0, 1]$*)

$$\theta 3x = \text{Table}[\text{Sum}[x(i-1)\Delta x\ ;$$

$$\text{Evaluate}\big[\text{hf3}\ \Delta\text{WListMatrix}\big[\big[i, j\big]\big]\big], \{i, 1, \text{N}\}\big], \{j, 1, 100\}\big];$$

$$\text{ListLinePlot}[\theta 3x, \text{Frame} \to \text{True}, \text{Axes} \to \text{True}, \text{Joined} \to \text{True},$$

$$\text{PlotLabel} \to \theta 3x\,\text{realizations}, \Delta x = 0.0001,$$

$$\text{AxesLabel} \to \{\text{HoldForm}[\text{Realization no}]\},$$

$$\text{ColorFunction} \to \text{Automatic},$$

$$\text{LabelStyle} \to \{\text{FontFamily} \to \text{"Times New Roman"}, 14, \text{GrayLevel}[0], \text{FontWeight} \to \text{Bold}\}\big]$$

$$\text{ListLinePlot}[\{\theta 1x, \theta 2x, \theta 3x\}, \text{Frame} \to \text{True}, \text{Axes} \to \text{True},$$

$$\text{Joined} \to \text{True},$$

$$\text{PlotLabel} \to \text{"}\theta 1x, \theta 2x, \theta 3x\,\text{realizations}, \Delta x = 0.0001\text{"},$$

$$\text{PlotLegends} \to \text{Placed}[\{\theta 1x, \theta 2x, \theta 3x\}, \{\text{Center}, \text{Below}\}],$$

$$\text{AxesLabel} \to \{\text{HoldForm}[\text{Realization no}]\},$$

$$\text{ColorFunction} \to \text{Automatic},$$

$$\text{LabelStyle} \to \{\text{FontFamily} \to \text{"Times New Roman"}, 14, \text{GrayLevel}[0], \text{FontWeight} \to \text{Bold}\}\big]$$

(*Compute $\theta_4 x$ realizations for Hermite function hf4; Each realization produces a single value over $x = [0, 1]$*)

$$\theta 4x = \text{Table}[\text{Sum}[x(i-1)\Delta x\,;$$

$$\text{Evaluate}\big[\text{hf4}\ \Delta\text{WListMatrix}[[i,j]]\big], \{i, 1, \text{N}\}\big], \{j, 1, 100\}\big];$$

$$\text{ListLinePlot}[\theta 4x, \text{Frame} \to \text{True}, \text{Axes} \to \text{True}, \text{Joined} \to \text{True},$$

$$\text{PlotLabel} \to "\theta 4x\,\text{realizations},\ \Delta x = 0.0001",$$

$$\text{AxesLabel} \to \{\text{HoldForm}[\text{Realization no}]\},$$

$$\text{ColorFunction} \to \text{Automatic},$$

$$\text{LabelStyle} \to \{\text{FontFamily} \to "\text{Times New Roman}",\ 14,\ \text{GrayLevel}[0],$$

$$\text{FontWeight} \to \text{Bold}\}]$$

(*Compute $\theta_5 x$ realizations for Hermite function hf5; Each realization produces a single value over $x = [0, 1]$*)

$$\theta 5x = \text{Table}[\text{Sum}[x(i-1)\Delta x\,;$$

$$\text{Evaluate}\big[\text{hf5}\ \Delta\text{WListMatrix}[[i,j]]\big], \{i, 1, \text{N}\}\big], \{j, 1, 100\}\big];$$

$$\text{ListLinePlot}[\theta 5x, \text{Frame} \to \text{True}, \text{Axes} \to \text{True}, \text{Joined} \to \text{True},$$

$$\text{PlotLabel} \to "\theta 5x\,\text{realizations},\ \Delta x = 0.0001",$$

$$\text{AxesLabel} \to \{\text{HoldForm}[\text{Realization no}]\},$$

$$\text{ColorFunction} \to \text{Automatic},$$

$$\text{LabelStyle} \to \{\text{FontFamily} \to "\text{Times New Roman}",\ 14,\ \text{GrayLevel}[0],$$

FontWeight → Bold}]

(*Compute $\theta_6 x$ realizations for Hermite function hf6; Each realization produces a single value over $x = [0, 1]$*)

```
θ6x = Table[Sum[x(i − 1)Δx ;

Evaluate[hf6 ΔWListMatrix[[i, j]]], {i, 1, N}], {j, 1, 100}];

ListLinePlot[θ6x, Frame → True, Axes → True, Joined → True,

PlotLabel → "θ6x realizations, Δx = 0.0001",

AxesLabel → {HoldForm[Realization no]},

ColorFunction → Automatic,

LabelStyle → {FontFamily
        → "Times New Roman", 14, GrayLevel[0],

FontWeight → Bold}]
```

(*Plot θ4x, θ5x, and θ6x together.*)

```
ListLinePlot[{θ4x, θ5x, θ6x}, Frame → True, Axes → True,

Joined → True, PlotRange → All,

PlotLabel → "θ4x, θ5x, θ6x realizations, Δx = 0.0001",

PlotLegends → Placed[{θ4x, θ5x, θ6x}, {Center, Below}],

AxesLabel → {HoldForm[Realization no]},

ColorFunction → Automatic,

LabelStyle → {FontFamily
```

→ "Times New Roman", 14, GrayLevel[0],

FontWeight → Bold}]

(*Computation of $H(3, 0, 2)$ realizations*)

$$\text{H302} = \left(\theta 1x^3 - 3\theta 1x\right)\left(\theta 3x^2 - 1\right);$$

ListLinePlot[H302, Frame → True, Axes → True,

Joined → True,

PlotRange → All, PlotLabel → "H302 realizations, $\Delta x = 0.0001$",

AxesLabel → {HoldForm[Realization no]},

ColorFunction → Automatic,

LabelStyle → {FontFamily
→ "Times New Roman", 14, GrayLevel[0],

FontWeight → Bold}]

(*Computation of $H(2, 0, 3)$ realizations*)

$$\text{H203} = \left(\theta 1x^2 - 1\right)\left(\theta 3x^3 - 3\theta 3x\right);$$

(*Plot H302 and H203 together *)

ListLinePlot[{H203, H302}, Frame → True, Axes → True,

Joined → True, PlotRange → All,

PlotLabel → "H203 and H302 realizations, $\Delta x = 0.0001$",

AxesLabel → {HoldForm[Realization no]},

PlotLegends → Placed[{H203, H302}, {Center, Below}],

ColorFunction → Automatic,

LabelStyle → {FontFamily
→ "Times New Roman", 14, GrayLevel[0],
FontWeight → Bold}]

Appendix 2.2: A Summary of Multi-index Notation

As we encounter multi-index notation often in the white noise analysis, a brief description is given here. N_0^n is defined as the n dimensional set of natural numbers.$\alpha = (\alpha_1, \alpha_2, \ldots, \alpha_n) \in N_0^n$ is an n-tuple of nonnegative integers. We define the following operations if $\alpha, \beta \in N_0^n$ and $\gamma = (\gamma_1, \gamma_2, \ldots, \gamma_n) \in \mathbb{R}^n$:

- $\alpha \pm \beta = (\alpha_1 \pm \beta_1, \alpha_2 \pm \beta_2, \ldots, \alpha_n \pm \beta_n)$.
- $\alpha \leq \beta \Leftrightarrow \alpha_i \leq \beta_i; \forall i \in \{1, 2, \ldots, n\}$.
- Absolute value is the sum of components: $|\alpha| = \alpha_1 + \alpha_2 + \cdots + \alpha_n$.
- Definition of factorial: $\alpha! = \alpha_1!\alpha_2! \ldots \alpha_n!$.
- Definition of binomial coefficient: $\binom{\alpha}{\beta} = \binom{\alpha_1}{\beta_1}\binom{\alpha_2}{\beta_2} \cdots \binom{\alpha_n}{\beta_n} = \frac{\alpha!}{\beta!(\alpha-\beta)!}$.
- Definition of multinormal coefficient: $\binom{k}{\alpha} = \frac{k!}{\alpha_1!\alpha_2!\ldots\alpha_n!} = \frac{k!}{\alpha!}$ where $k := |\alpha| \in N_0$.
- Definition of powers: $x^\alpha = x_1^{\alpha_1} x_2^{\alpha_2} \ldots x_3^{\alpha_n}$.
- Definitions of partial derivatives: $\partial^\alpha = \partial_1^{\alpha_1} \partial_2^{\alpha_2} \ldots \partial_n^{\alpha_n}$, where $\partial_i^{\alpha_i} := \frac{\partial^{\alpha_i}}{\partial x_i^{\alpha_i}} \equiv D^\alpha$.

References

Debnath, L., & Mikusinski, P. (2005). *Introduction to Hilbert spaces with applications* (3rd ed.). Elsevier Academic Press.

Di Nunno, G., Øksendal, B., & Proske, F. (2019). *Malliavin calculus for Lévy processes with applications to finance*. Springer-Verlag.

Egge, E. S. (2019). *An introduction to symmetric functions and their combinatorics*. American Mathematical Society.

Hida, T. (1980). *Brownian motion*. Springer-Verlag.

Holden, H., & Øksendal, B. (2022). A white noise approach to stochastic differential equations driven by Wiener and Poisson processes. In *Nonlinear theory of generalized functions*. https://doi.org/10.1201/9780203745458-28

Holden, H., et al. (1996). *Stochastic partial differential equations*. Birkhäuser Boston. https://doi.org/10.1007/978-1-4684-9215-6

Holden, H., et al. (2010). *Stochastic partial differential equations*. Springer New York. https://doi.org/10.1007/978-0-387-89488-1
Ito, K. (1951). Multiple Wiener integral. *Journal of the Mathematical Society of Japan, 3*(1), 157–169. https://doi.org/10.2969/JMSJ/00310157
Kulasiri, D., & Kosarwal, R. (2021). Chemical master equation for large biological networks. In *State-space expansion methods using AI*. https://doi.org/10.1007/978-981-16-5351-3
Øksendal, B. (1998). *Stochastic differential equations: An introduction with applications* (5th ed.). Universitext.
Wiener, N. (1938). The homogeneous chaos. *American Journal of Mathematics, 60*(4), 897. https://doi.org/10.2307/2371268

Chapter 3
Wick Algebra, Diffusions, and Computations

3.1 Introduction

We summarize pertinent mathematical results in Chap. 2, and in this chapter, we focus on more useful derivations based on the Wick products with the aim of developing a consistent mathematical machinery for computations. Central to these developments are the Hermite polynomials, and they are used in the Wick products related to the Itô diffusions. The idea of using the Itô diffusions to model the stochastic variables that participate in the Wick type equations is inspired by the fact that many physical variables have drift and diffusive components to them. The forces and flows governed by the physical laws drive the average behaviors in noisy environments and the fluctuations in intrinsic and extrinsic forces and flows give rise to the "noise" in the system. Therefore, it is reasonable to assume that the drift term represents the average behavior, and the diffusion term is a result of the fluctuations. Another important assumption is that the effects of average behaviors and the fluctuations are additive, not multiplicative, and to simplify the analysis, we assume the coefficients associated with these terms are constants.

In our formulations, stochasticity or "noise" is driven by the diffusive component of a variable, which is expressed as an Itô integral with the Wiener process embedded. In Chap. 2, we mention that the Wick calculus is like normal calculus of deterministic variables rendering the use of the Itô calculus unnecessary. However, the Wick calculus has some noted idiosyncrasies as we observe later, but it has numerical convergence properties that help in computations, mainly due to the Hermite functions, which belong to the Schwartz space of rapidly decreasing smooth functions. Our intention is to provide the mathematical formulas without proof as we have done before, and we follow (Nunno et al., 2019) closely in these formulas and notations.

D. Kulasiri, *Stochastic Differential Equations for Chemical Transformations in White Noise Probability Space*, https://doi.org/10.1007/978-981-97-9392-1_3

3.2 Useful Properties of the Wick Products and I_n

We consider that X, Y, $Z \in (S^*)$, and the Wick products of these stochastic variables follow commutative, associative, and distributive laws. We define the Wick powers as follows:

$$X^{\diamond n} = X \diamond X \diamond \ldots \diamond X,\ n = 1, 2, \ldots,\ \text{and } X^{\diamond 0} = 1.$$

The Wick exponential of X is defined similar to the deterministic counterpart as,

$$\exp^{\diamond} X = \sum_{n=0}^{\infty} \frac{1}{n!} X^{\diamond n}, \tag{3.1}$$

if convergent in (S^*).

As mentioned previously, the Wick products obey algebraic rules similar to deterministic variables with some exceptions. For example,

$$(X + Y)^{\diamond 2} = X^{\diamond 2} + 2X \diamond Y + Y^{\diamond 2}, \tag{3.2}$$

and

$$\exp^{\diamond}(X + Y) = \exp^{\diamond b}(X) \cdot \exp^{\diamond}(Y). \tag{3.3}$$

But in general,

$$X \cdot (Y \diamond Z) \neq (X \cdot Y) \diamond Z, \text{ for example.}$$

When we mix ordinary products with the Wick products, we need to be cautious. If one of the variables is deterministic, then the Wick products and the normal products are the same, which is a very useful result in computations.

Another useful relationship is related to the expectation of the Wick product:

$$E[X \diamond Y] = E[X] \cdot E[Y]. \tag{3.4}$$

The assumption here is that X, Y, and X$\diamond$Y are integrable in $L^2(P)$ space. Equation (3.4) does not mean however that X and Y are independent variables. From Eq. (3.4), it can be shown that

$$E\left[\exp^{\diamond} X\right] = \exp[E[X]]. \tag{3.5}$$

We apply Eq. (3.5) quite regularly in applications to obtain the mean values of an expression containing the Wick product. Another important result we use in our computations is the following:

If $0 \leq t_1 \leq t_2 \leq t_3 \leq t_4$, then

$$(W(t_4) - W(t_3))\diamond(W(t_2) - W(t_1))$$
$$= (W(t_4) - W(t_3)) \cdot (W(t_2) - W(t_1)). \tag{3.6}$$

Equation (3.6) can be modified, if $t_1 = 1\Delta t$; $t_2 = 2\Delta t$; $t_3 = 3\Delta t$; $t_4 = 4\Delta t$; and $\Delta t > 0$, then.

$$(W(4\Delta t) - W(3\Delta t))\diamond(W(2\Delta t) - W(1\Delta t))$$
$$= (W(4\Delta t) - W(3\Delta t)) \cdot (W(2\Delta t) - W(1\Delta t)). \tag{3.7}$$

This can be generalized for an interval $\Delta t \le t \le n\Delta t$, starting with i th equidistant interval Δt:

$$(W(i+3)\Delta t) - W((i+2)\Delta t)\diamond(W(i+1)\Delta t) - W((i\Delta t))$$
$$= (W(i+3)\Delta t) - W((i+2)\Delta t) \cdot (W(i+1)\Delta t) - W((i\Delta t)) \tag{3.8}$$

Eq. (3.8) is a very useful result in numerical computations.

For a stochastic $\mathcal{F}_t$-measurable function F, the following relationship is useful:

$$F\diamond(W(t+\Delta t) - W(t)) = F \cdot (W(t+\Delta t) - W(t)). \tag{3.9}$$

There is a chain rule for a function f, ($f(\mathbb{R}) \in \mathbb{R}$), which is valid in (S*):

$$\frac{df^{\diamond}(X(t))}{dt} = (f')^{\diamond}(X(t))\diamond\frac{d(X(t))}{dt}. \tag{3.10}$$

For example, if

$f(t) = \exp(at)$, $a \in \mathbb{R}$ and is a constant, then

$$\frac{d\ \exp^{\diamond}(aX(t))}{dt} = (a\ \exp(a\ t))^{\diamond}(X(t))\diamond\frac{dX(t)}{dt},$$
$$= a\ \exp(a\ t)^{\diamond}(X(t))\diamond\frac{dX(t)}{dt}. \tag{3.11}$$

For the deterministic functions, $f(x)$, $g(y) \in L^2(\mathbb{R})$, it can be shown that (See (Nunno et al., 2019)), within limits in $\mathbb{R}$,

$$\left(\int f(x)dW(x)\right)\diamond\left(\int g(y)dW(y)\right) = \left(\int f(x)dW(x)\right)$$
$$\left(\int g(y)dW(y)\right) - \int f(t)g(t)dt. \tag{3.12}$$

From Eq. (3.12), the following equation can be derived that is quite useful in computations:

$$g^{\diamond n} = \|g\|^n h_n\left(\frac{\theta}{\|g\|}\right) \tag{3.13}$$

$$\text{with } \theta = \int dW(t). \tag{3.14}$$

Here $g \in \mathbb{R}$.

For example, if $g = 1, \theta = W(t)$, then, $\|g\|^2 = \int_0^t dt, \|g\| = \sqrt{t}$. Therefore using Eq. (3.13),

$$\mathbf{W}^{\diamond n} = \boldsymbol{t}^{n/2}\boldsymbol{h}_n\left(\frac{\mathbf{W}(\boldsymbol{t})}{\sqrt{\boldsymbol{t}}}\right). \tag{3.15}$$

Using the Hermite polynomials, $W^{\diamond 0}(t) = 1$; $W^{\diamond 0}(t) = 1$;

$$\mathbf{W}^{\diamond 1}(\boldsymbol{t}) = \mathbf{W}(\boldsymbol{t});$$

$$\mathbf{W}^{\diamond 2}(\boldsymbol{t}) = \mathbf{W}^2(\boldsymbol{t}) - \boldsymbol{t}; \text{ and}$$

$$W^{\diamond 3}(t) = W^3(t) - 3tW(t).$$

We can see that from Chap. 2,

$$I_n(1) = W^{\diamond n}(t).$$

It can be shown that,

$$\exp^{\diamond}\left\{\int_0^t g\, dW\right\} = \exp\left\{\int_0^t g\, dW - \frac{1}{2}\|g\|^2\right\}, g(y) \in L^2(\mathbb{R}). \tag{3.16}$$

Taking $g = 1$, one can show that,

$$\exp^{\diamond}[W(t)] = \exp\left[W(t) - \frac{1}{2}t\right], t \geq 0. \tag{3.17}$$

This shows that the Wick exponent of $W(t)$ is mediated by time as time increases.

There is an interesting relationship between the Itô iterated integral $I_n(g)$ as defined in Chap. 2 and the Hermite polynomials similar to the relationships given in Eq. (3.13). Recall that,

$$I_n(g) = \|g\|^n h_n\left(\frac{\theta}{\|g\|_{L^2([0,T])}}\right),$$

and we can make use of this equation to show that, if $g = 1$,

$$I_n(1) = t^{(n/2)} h_n\left(\frac{W(t)}{\sqrt{t}}\right). \tag{3.18}$$

From this we can see that,
$W^{\diamond n} = I_n(1)$, and also,

$$I_0(1) = 1,$$

$$I_1(1) = W(t),$$

$$I_2(1) = W^2(t) - t,$$

$$I_3(1) = W^3(t) - 3tW(t),$$

$$I_4(1) = W^4(t) - 6tW^2(t) + 3t^2,$$

$$I_5(1) = W^5(1) - 10tW^3(t) + 15, \ldots \text{so on.}$$

We can use these equations to compute $W^{(\diamond n)}$ for a given n.

3.3 The Wick Algebra for Itô Diffusions

In applications, it is reasonable to assume that stochastic variables behave similar to the behavior of Itô diffusions (Klebaner, 2012) with a drift term and a diffusion term that are additive, both dependent on time, t. Hence, we can define,

$$X(t) = \int_0^t \mu dt + \int_0^t \sigma dt. \tag{3.19}$$

where μ and σ are deterministic functions of t or constants. In our treatment here, we assume that they are constants to simplify mathematics.

Let us define

$$\alpha \equiv \int_0^t \mu dt, \text{ and} \tag{3.20}$$

$$\theta \equiv \int_0^t \sigma \, dW. \tag{3.21}$$

Here θ has the same form as the previously used θ but have different meanings: θ is a diffusion term that models the fluctuations in a quantity of interest. Then,

$$X(t) = \alpha + \theta. \tag{3.22}$$

We now express, $X^{(\Diamond 2)} = \alpha^{(\Diamond 2)} + 2\alpha\Diamond\theta + \theta^{(\Diamond 2)}$, and as α is either a deterministic function of t or constant, $\alpha\Diamond\theta = \alpha\theta$; i.e., the Wick product is valid only between stochastic variables.

$$\begin{aligned} X(t)^{\Diamond 2} &= \alpha^2 + 2\alpha\theta + \|\sigma\|^2 \left[h_2 \left(\frac{\theta}{\|\sigma\|} \right)^2 \right] \\ &= \alpha^2 + 2\alpha\theta + \theta^2 - \|\sigma\|^2, \end{aligned} \tag{3.23}$$

where $\|\sigma\|^2 = \int_0^t \sigma^2 dt = \sigma^2 t$, and because we assume σ to be a constant.

$$\begin{aligned} X(t)^{\Diamond 2} &= (\alpha^2 + 2\alpha\theta + \theta^2) - \sigma^2 t \text{ and} \\ X(t)^{\Diamond 2} &= X(t)^2 - \sigma^2 t \end{aligned} \tag{3.24}$$

This shows that the Wick square of an Itô diffusion decreases with time and in certain realizations, it could be negative as well for large t, especially when the diffusion coefficient is large.

Suppose $\alpha_i = \int_0^t \mu_i dt$; $\theta_i = \int_0^t \sigma_i dW(t)$, $(i = 1, 2, 3)$.

$$\begin{aligned} \text{Then, } X_1\Diamond X_2 &= (\alpha_1 + \theta_1)\Diamond(\alpha_2 + \theta_2), \\ &= \alpha_1\alpha_2 + \alpha_1\theta_2 + \alpha_2\theta_1 + \theta_1\Diamond\theta_2. \end{aligned}$$

From Eq. (3.21),

$$\begin{aligned} \theta_1\Diamond\theta_2 &= (\int \sigma_1 W(t))\Diamond(\int \sigma_2 dW(t)), \\ &= \theta_1\theta_2 - \int \sigma_1\sigma_2 dt, \\ &= \theta_1\theta_2 - \beta_{12}, \end{aligned}$$

where $\beta_{ij} \equiv \int \sigma_i\sigma_j dt$

$$\therefore X_1\Diamond X_2 = \alpha_1\alpha_2 + \alpha_1\theta_2 + \alpha_2\theta_1 + \theta_1\theta_2 - \beta_{12}. \tag{3.35}$$

This equation can also be written as,

$$X_1 \Diamond X_2 = X_1 \cdot X_2 - \sigma_1 \sigma_2 t.$$

Extending Eq. (3.35):

$$i = 1, 2, \quad X_i \Diamond X_j = \alpha_i \alpha_j + \alpha_i \theta_j + \alpha_j \theta_i + \theta_i \theta_j - \beta_{ij}. \tag{3.36}$$

We can interchange i and j and see that the Wick product of two Itô diffusions are associative. By putting $i = j$, we can obtain the expression for $X(t)^{\Diamond 2}$ that we derive in Eq. (3.34).

Let us derive the expression for $X_1 \Diamond X_2 \Diamond X_3$.

$$X_1 \Diamond X_2 \Diamond X_3 = (X_1 \Diamond X_2) \Diamond X_3.$$

Substituting Eq. (3.35),

$$\begin{aligned}
&(\alpha_1\alpha_2 + \alpha_1\theta_2 + \alpha_2\theta_1 + \theta_1\theta_2 - \beta_{12}) \Diamond (\alpha_3 + \theta_3) \\
&= (\alpha_1\alpha_2\alpha_3 + \alpha_1\alpha_3\theta_2 + \alpha_2\alpha_3\theta_1 + \alpha_3\theta_1\theta_2 - \alpha_3\beta_{12}) \\
&\quad + (\alpha_1\alpha_2\theta_3 + \alpha_1\theta_2\theta_3 + \alpha_2\theta_1 \Diamond \theta_3 + \theta_1\theta_2 \Diamond \theta_3 - \beta_{12}\theta_3),
\end{aligned}$$

$$\begin{aligned}
X_1 \Diamond X_2 \Diamond X_3 &= \alpha_1\alpha_2\alpha_3 + \alpha_1\alpha_2\theta_3 + \alpha_2\alpha_3\theta_1 + \alpha_1\alpha_3\theta_2 \\
&\quad + \alpha_1\theta_2\theta_3 + \alpha_3\theta_1\theta_2 + \alpha_2\theta_1 \Diamond \theta_3 + \theta_1\theta_2 \Diamond \theta_3 \\
&\quad - \alpha_3\beta_{12} - \theta_3\beta_{12}.
\end{aligned}$$

In numerical calculations, we use Eq. (3.35) to compute the two terms Wick products and evaluate the expression $X_1 \Diamond X_2 \Diamond X_3$.

It is often necessary to compute the Wick exponentials of Itô diffusions. To that end the following relationship is quite useful for the diffusion term. (See (Nunno et al., 2019))

$$\exp^{\Diamond}\{\textstyle\int g dW\} = \exp\left\{\int g dW - \frac{1}{2}\|g\|^2\right\}, \quad g \in L^2(\mathbb{R}).$$

Taking $g = 1$, one can show that,

$$\exp^{\Diamond}\{W(t)\} = \exp\left\{W(t) - \frac{1}{2}t\right\}, t \leq 0. \tag{3.37}$$

Equation (3.37) is quite useful in computing the Wick exponential of the Wiener process as we see later.

Now, we can calculate the Wick exponential of an Itô diffusion, X(t), as defined before.

$$\exp^{\Diamond}[X(t)] = \exp^{\Diamond}[\alpha + \theta],$$

$$= \exp^{\diamond}[\alpha]\diamond \exp^{\diamond}[\theta]. \quad (3.38)$$

As α is a deterministic function of t,

$$\exp^{\diamond}[\mathrm{X(t)}] = \exp[\alpha]\exp\left\{\int \sigma\ \mathrm{dW} - \frac{1}{2}\|\sigma\|^2\right\},$$

$$= \exp[\mu\ \mathrm{t}]\exp\left\{\sigma\ \mathrm{W} - \frac{1}{2}\sigma^2\mathrm{t}\right\}$$

$$= \exp[\mu\ t + \sigma\ \mathrm{W}]\exp\left\{-\frac{1}{2}\sigma^2 t\right\},$$

$$\therefore \exp^{\circ}[X(t)] = \exp(X)\exp\left\{-\frac{1}{2}\sigma^2 t\right\}. \quad (3.39)$$

We use Eq. (3.39) in our computations in the next section.

3.4 Computations

It often helps to understand the nature of the Wick products through computer simulations. In this section, we compute the quantities we derive in the previous section and in Chap. 2 using Mathematica® (version 14.0), and the Mathematica® program is given in the Appendix 3.1.

For computational purposes, we assume $\mathbb{R} = [0, 1]$ and $\Delta t = 10^{-5}$ time units. In later chapters, we specify the units, but in this section, we only focus on the behaviors of the computed quantities. As all these quantities are driven by the Wiener process, it is important to keep the same Wiener process for the quantities that are used for comparison. As we see in the Mathematica® program, the behaviors of various quantities at individual realization level are heavily influenced by the Wiener process, hence a large number of realizations are needed to obtain to interpret the phenomenon at hand. Here it is important for us to resort to statistical properties of the realizations. We compare two different Itô diffusions, one with smaller drift and diffusion coefficients and the other with larger coefficients. As we mentioned before, we assume that these coefficients are constants which are quite reasonable in practical situations. For$X_1(t)$, $\mu_1 = 0.2, \sigma_1 = 0.2$; for$X_2(t)$, $\mu_2 = 0.6, \sigma_2 = 0.6$. As mentioned previously, we use the same Wiener realization that is shown in Fig. 3.1 to compute the realizations of X_1 (Fig. 3.2) and X_2 (Fig. 3.3). Figure 3.4 and Fig. 3.5 show the corresponding realizations of $X_1 \cdot X_2$ and $X_1 \diamond X_2$, respectively. It is interesting to note that the influence of Wiener realization used is quite dominant in the trends or patterns of X_1 and X_2 realizations. However, $X_1 \cdot X_2$ and $X_1 \diamond X_2$ are distinctly different in the sense that the Wick product produces the realization smaller than the normal product and it moves in the negative region when the normal product is positive. In another words, when two positive stochastic variables are multiplied in the normal manner, one would expect a positive outcome, but the Wick multiplication renders a negative outcome during certain periods. This is quite counterintuitive! It should be noted that the Wiener realization we use moves in the negative region for a long period of time, and high diffusion coefficients we deliberately use to highlight this effect. This also means that when two variables have very high noise, the Wick multiplication would turn negative at some periods in time depending on the Wiener realization. This shows that the effects of randomness have more profound implications if we use the Wick products.

Given the realization in Fig. 3.1, the Itô diffusion X_1 with the drift coefficient 0.2 and the diffusion coefficient 0.2, is expected to traverse a path similar the Wiener process realization but drifting toward to the right but the diffusion coefficient tends to reduce the absolute values of the amplitude of X_1 (Fig. 3.2). We can think of Wiener process itself as an Itô diffusion with zero drift and the diffusion coefficient of 1.

With the same line of reasoning, we see that the diffusion X_2 (Fig. 3.2) drifting more toward to the right compared to the Wiener realization with numerically higher amplitude (absolute values) at any given t compared to those of X_1.

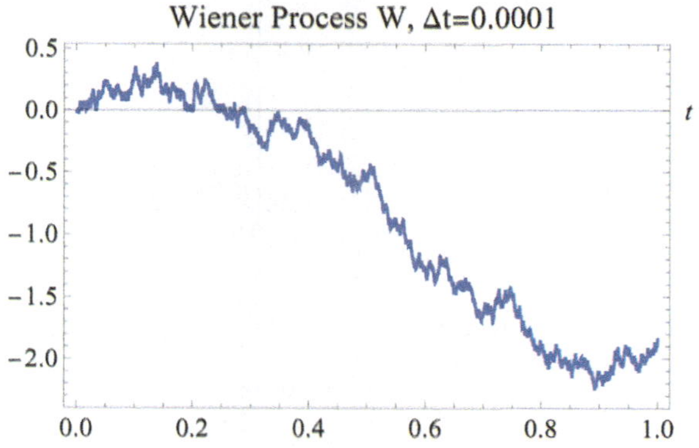

Fig. 3.1 Wiener process realization used in the computation of X_1, X_2, $X_1 \cdot X_2$, and $X_1 \diamond X_2$. Note the movement into the negative region

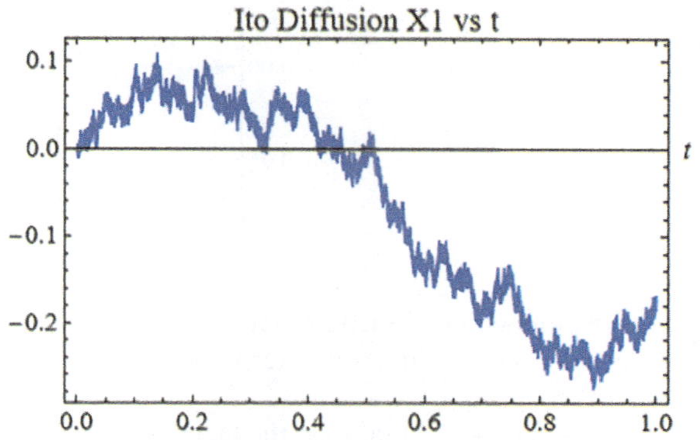

Fig. 3.2 Realization of X_1 for the Wiener realization given in Fig. 3.1

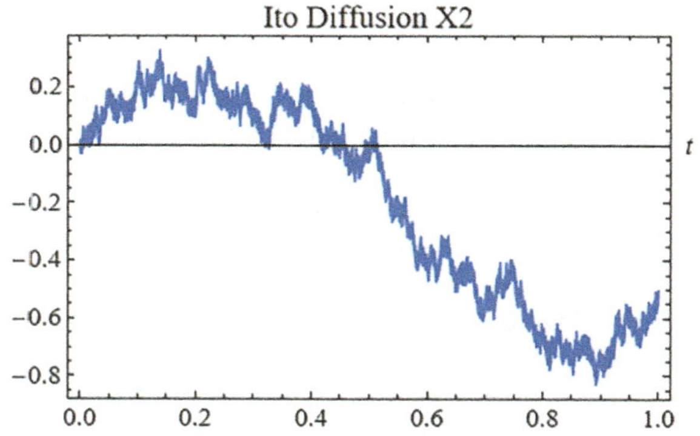

Fig. 3.3 Realization of X_2 for the Wiener realization given in Fig. 3.1

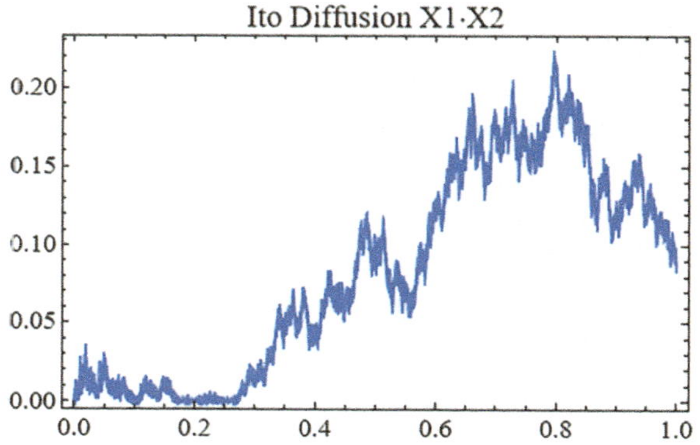

Fig. 3.4 Realization of $X_1 \cdot X_2$

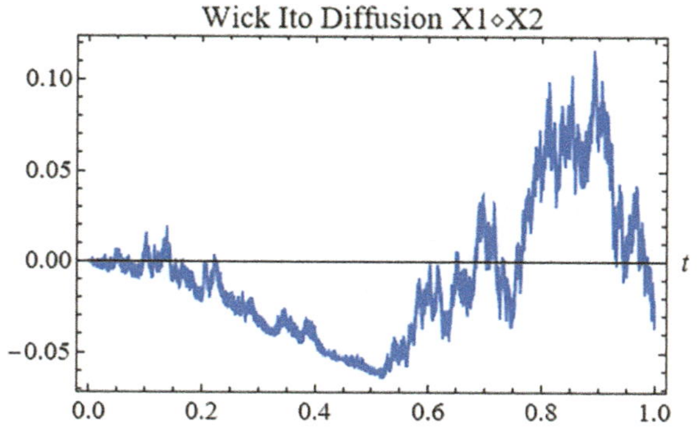

Fig. 3.5 Realization of $X_1 \Diamond X_2$

As mentioned previously, $X_1 \cdot X_2$ and $X_1 \Diamond X_2$ are quite distinct random processes which can lead to different interpretations when used in modeling variables in complex systems.

Figures 3.4 and 3.5 highlight the fact that even though $X_1 \cdot X_2$ moves in the positive region, $X_1 \Diamond X_2$ could be negative; this is not necessary the case in many realizations

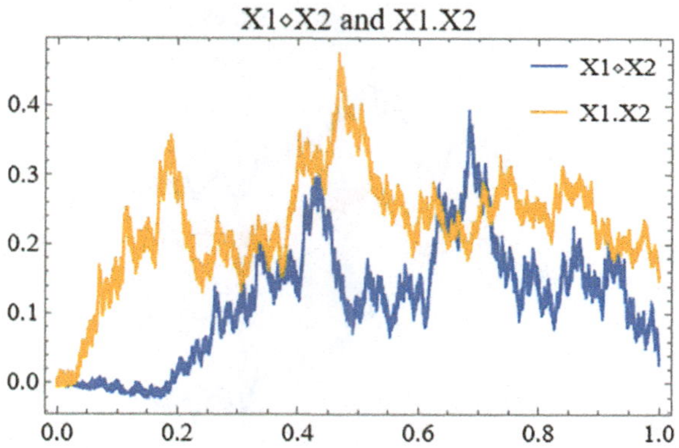

Fig. 3.6 Realizations of $X_1 \cdot X_2$ and $X_1 \diamond X_2$ that are mostly in positive region

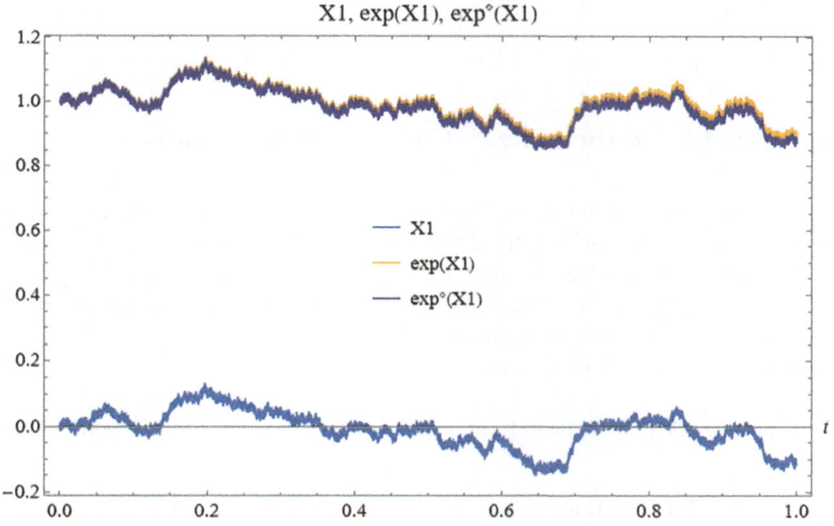

Fig. 3.7 Realizations of X_1, exp (X_1), and exp $\diamond(X_1)$ for the same Wiener process

as we see when many realizations are generated. An example of both $X_1 \cdot X_2$ and $X_1 \diamond X_2$ mostly in the positive region is given in Fig. 3.6.

We compute the normal and Wick exponentials of X_1 and X_2 next.

Figure 3.7 shows that the exponentials, both normal and Wick, are larger than X_1 and both exponentials move close to each other over time, and normal exponential is slightly larger than the Wick exponential.

According to Fig. 3.8, for larger values of drift (0.6) and diffusion coefficient (0.6), the Wick exponential stays larger than the normal exponential and they converge to each other for larger t values.

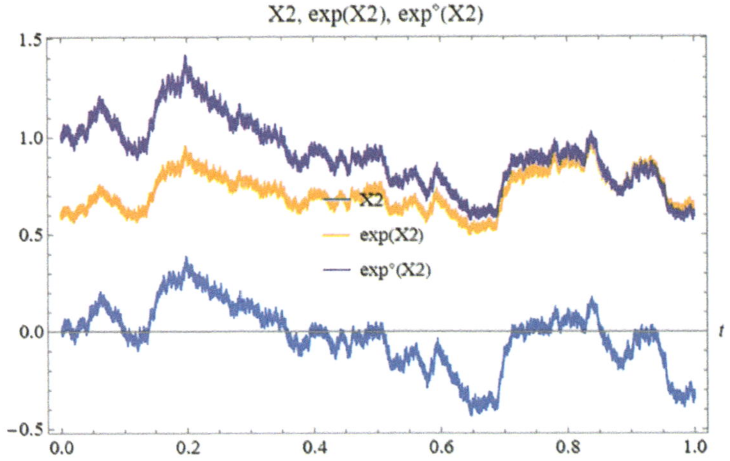

Fig. 3.8 Realizations of X_2, exp (X_2), and exp $\diamond(X_2)$ for the same Wiener process used in Fig. 3.7

Appendix 3.1: Mathematica® Code Without Outputs

(*We evaluate the computations involving the Wick product; first we compute X1◇X1 verses X1·X2 of two Ito diffusions defined as dX = μ_i dt + σ_i dW, (i = 1, 2), then we change the drift and diffusion coefficients to see the differences in the product. We use the time interval [0, 1] with Δt = 0.0001 to make sure that all the Wiener increments are small enough. *)

(∗Module for Wiener process∗)

```
Δt = 0.0001; N = 10000; μ1 = 0.2; σ1 = 0.2; μ2 = 0.6; σ2 = 0.6;

ΔWList = Table[N[Random[NormalDistribution[0.0, Sqrt[Δt]]], 32], {N}];
```

(*Create a Wiener increment list with time*)

```
ΔWtList = Table[{(i − 1)Δt, ΔWList[[i]]}, {i, 1, N}];
```

(*Plot this to see the Wiener increments*)

```
ListLinePlot[ΔWtList, Frame → True, Axes → True,
Joined → True, PlotLabel → Wiener Increments ΔW,
AxesLabel → {HoldForm[t], HoldForm[ΔW]},
ColorFunction → Automatic,
LabelStyle− > {FontFamily
```

```
→ "Times New Roman", 14, GrayLevel[0]}]

(*Compute X1 and X2 with μi and σi;
WP = Wiener process ; WPtList is the Winer process  versus time*)

WP = Table[Sum[ΔWList[[i]], {i, 1, i}], {i, 1, N}];
WPtList = Table[{(i − 1)Δt, WP[[i]]}, {i, 1, N}];

(*Plot the Wiener process*)

ListLinePlot[WPtList, Frame → True, Axes → True,
Joined → True, PlotLabel → Wiener Process W, Δt = 0.0001,
AxesLabel → {HoldForm[t]}, ColorFunction → Automatic,
  LabelStyle− > {FontFamily
       → "Times New Roman", 14, GrayLevel[0]}]

(*Compute X1*)

X1 = Table[μ1Δt(i − 1) + σ1WP[[i]], {i, 1, N}];
X1tList = Table[{(i − 1)Δt, X1[[i]]}, {i, 1, N}];

(*Plot X1 versus t *)

ListLinePlot[X1tList, Frame → True, Axes → True,
Joined → True, PlotLabel → Ito Diffusion X1 versus t,
AxesLabel → {HoldForm[t]},
ColorFunction → Automatic,
LabelStyle− > {FontFamily
       → "Times New Roman", 14, GrayLevel[0]}]

(*Compute X2 *)

X2 = Table[μ2Δt(i − 1) + σ2WP[[i]], {i, 1, N}];
X2tList = Table[{(i − 1)Δt, X2[[i]]}, {i, 1, N}];

(*Plot X2 *)

ListLinePlot[X2tList, Frame → True, Axes → True,
Joined → True, PlotLabel → Ito Diffusion X2,
AxesLabel → {HoldForm[t]},
```

```
ColorFunction → Automatic,
LabelStyle− > {FontFamily
        → ″Times New Roman″, 14, GrayLevel[0]}]
```

(*Plot X1·X2 *)

```
ListLinePlot[X1tListX2tList, Frame → True, Axes → True,
Joined → True, PlotLabel → Ito Diffusion X1 · X2,
AxesLabel → {HoldForm[t]}, ColorFunction → Automatic,
LabelStyle− > {FontFamily
        → ″Times New Roman″, 14, GrayLevel[0]}]
```

(*Compute X1◇X2 versus *t* *)

```
X1◇X2 = Table[(X1X2)[[i]] − σ1σ2(i − 1)Δt, {i, 1, N}];
X1WickX2tList = Table[{(i − 1)Δt, (X1◇X2)[[i]]}, {i, 1, N}];
```

(* Plot X1◇X2 versus *t* *)

```
ListLinePlot[X1WickX2tList, Frame → True, Axes → True,
Joined → True, PlotLabel → Wick Ito Diffusion X1◇X2,
AxesLabel → {HoldForm[t]}, ColorFunction → Automatic,
LabelStyle− > {FontFamily
        → ″Times New Roman″, 14, GrayLevel[0]}]
```

(*Plot X1·X2 and X1◇X2 together *)

```
ListLinePlot[{X1WickX2tList, X1tListX2tList},
PlotLabel → X1◇X2 and X1.X2, Frame → True, Axes → True,
PlotRange → All,
PlotLegends → Placed[{X1◇X2, X1.X2}, {Right, Top}],
LabelStyle → {FontFamily
        → Times New Roman, 14, GrayLevel[0]},
ColorFunction− > Automatic]
```

(* We compute exponentials of X1 and X2, first the normal exponential and then the Wick exponential *)

```
Δt = 0.0001; N = 10000; μ1 = 0.2; σ1 = 0.2; μ2 = 0.6; σ2 = 0.6;
```

```
ΔWList = Table[N[Random[NormalDistribution[0.0, Sqrt[Δt]]], 32], {N}];
(*Create lists versus time as before*)

        ΔWtList = Table[{(i − 1)Δt, ΔWList[[i]]}, {i, 1, N}];
        WP = Table[Sum[ΔWList[[i]], {i, 1, i}], {i, 1, N}];
        WPtList = Table[{(i − 1)Δt, WP[[i]]}, {i, 1, N}];
        X1 = Table[μ1Δt(i − 1) + σ1WP[[i]], {i, 1, N}];
        X1tList = Table[{(i − 1)Δt, X1[[i]]}, {i, 1, N}];
        X2 = Table[μ2Δt(i − 1) + σ2WP[[i]], {i, 1, N}];
        X2tList = Table[{(i − 1)Δt, X2[[i]]}, {i, 1, N}];

(*Compute the exponentials *)

    expX1 = Table[Exp[μ1Δt(i − 1)]Exp[σ1WP[[i]]], {i, 1, N}];
    expX2 = Table[μ2Exp[Δt(i − 1)]Exp[σ2WP[[i]]], {i, 1, N}];

     expWickX1
      = Table[Exp[ − σ1^2Δt(i − 1)/2.]Exp[X1[[i]]], {i, 1, N}];
     expWickX2
      = Table[Exp[ − σ2^2Δt(i − 1)/2.]Exp[X2[[i]]], {i, 1, N}];

   expX1tList = Table[{(i − 1)Δt, expX1[[i]]}, {i, 1, N}];
   expWickX1tList = Table[{(i − 1)Δt, expWickX1[[i]]}, {i, 1, N}];
   expX2tList = Table[{(i − 1)Δt, expX2[[i]]}, {i, 1, N}];
   expWickX2tList = Table[{(i − 1)Δt, expWickX2[[i]]}, {i, 1, N}];

(* Plot X1 and its exponentials *)

  ListLinePlot[{X1tList, expX1tList, expWickX1tList},
  PlotLabel → X1, exp(X1),exp◇(X1), Frame → True,
  Axes → True, PlotRange → All, AxesLabel → {HoldForm[t]},
  PlotLegends → Placed[{X1, exp(X1), exp◇(X1)}, {Center, Center}],
  LabelStyle → {FontFamily
            → Times New Roman, 14, GrayLevel[0]},
            ColorFunction− > Automatic]

(* Plot X2 and its exponentials *)

  ListLinePlot[{X2tList, expX2tList, expWickX2tList},
```

PlotLabel → X2, exp(X2),exp$^\diamond$(X2), Frame → True,

Axes → True, PlotRange → All,

AxesLabel → {HoldForm[t]},

PlotLegends → Placed[{X2, exp(X2), exp$^\diamond$(X2)}, {Center, Center}],

LabelStyle → {FontFamily

→ Times New Roman, 14, GrayLevel[0]},

ColorFunction− > Automatic]

References

Klebaner, F. C. (2012). *Introduction to stochastic calculus with applications*, 3rd edn. Available at: https://doi.org/10.1142/P821.

Di Nunno, G., Øksendal, B., & Proske, F. (2019). *Malliavin calculus for Lévy processes with applications to finance*. Springer.

Chapter 4
Wick Stochastic Differential Equations

4.1 Introduction

The Itô integration is for the stochastic processes that are nonanticipating (adapted to the filtration $\mathbb{F}$). A nonanticipating stochastic variable has a value at a given time for a realization. Mathematically a stochastic variable X is nonanticipating (adapted) to $\mathbb{F}$ if X at a given time is a measurable function in a defined measurable space. It is predictable if the value at the next time is measurable with respect to σ-algebra related to $\mathbb{F}$.

The Itô integral requires the integrand to be adapted (or nonanticipating) to the Wiener (Brownian) filtration, $\mathcal{F}_t = \sigma(W_s, 0 \leq s \leq t),\ t \geq 0$. There are many situations where the integrand contains stochastic functions that are anticipating. The Itô integral can be expanded to include the processes that anticipate the Wiener processes. In our previous chapter, we discuss stochastic variables as the Itô diffusions, and if we assume that the variables in a model behave like Itô diffusions, then we anticipate other functions to be Itô diffusions as well. Strictly speaking this may not be the case in every situation but this is a "close enough" assumption. One example according to Itô is,

$$\int_0^1 W(1)\mathrm{d}W(s), \quad 0 \leq t \leq 1.$$

Here, $W(1)$ is an anticipating process. Another situation is when the initial conditions of an Itô SDE is anticipating. The Skorohod integral briefly discussed in Chap. 2 was developed to integrate possibly anticipating stochastic processes. If $X(t)$ is an anticipating process, then the Skorohod integral has an important property on $\mathbb{R}$,

$$\int X(t)\delta W(t) = \int X(t) \diamond w(t)\mathrm{d}t, \tag{4.1}$$

D. Kulasiri, *Stochastic Differential Equations for Chemical Transformations in White Noise Probability Space*, https://doi.org/10.1007/978-981-97-9392-1_4

here $w(t)$ is the white noise process. The right hand of this equation is an (S^*)-valued integral and if $X(t, \omega)$ is nonanticipating with respect to filtration $\mathbb{F}$, the Skorohod integral is the same as the Itô integral. In situations where we are not certain of the nature of stochastic processes involved in terms of anticipating and nonanticipating, it is safe to use the Skorohod integrals in stochastic differential equations, which then turns into the Wick products of the process and the white noise using the property given in Eq. (4.1).

Another significant advantage using WSDEs is the fact that, as mentioned in Chap. 2 already, the Wick products and their integrals can be manipulated using normal algebra and calculus without resorting to Itô calculus. In addition, the Wick product computations are based on the Hermite polynomials and functions, which are easily computable to very good accuracies provided that the time intervals used are very small. WSDEs can be solved analytically based on the work reported in Holden et al. (2010) and Di Nunno et al. (2019) using Hermite transforms, and in this chapter, we introduce the analytical solutions, where solutions are analytically available. In the next sections, we briefly discuss the theorems and lemmas that are pertinent to the Skorohod integration and WSDEs. In the last section, we present a model for positive noise for the coefficients involved in a SDE that may be useful in modeling physical systems to prevent the dependent variable moving to negative values where it is unrealistic to happen, for example, if the variable concerned is concentration of a substance in water, which must be nonnegative.

4.2 The Hermite Transform and WSDEs

Lindstrøm et al. (1991) first developed the Hermite transform that is used to solve WSDEs in this chapter. Essentially the Hermite transform, also called $\mathcal{H}$-transform, converts the stochastic variable, $X \in (S)_{-1}$, into the deterministic functions of complex variables $z_j \in \mathbb{C}, j = 1, 2, \ldots$. $(S)_{-1}$ is the Kondratiev distribution space, and $(S)_{-1} \supseteq (S^*)$. The $\mathcal{H}$-transform of

$$X = \sum_{\alpha} c_\alpha H_\alpha \tag{4.2}$$

is written as $\mathcal{H}X$ or $\tilde{X}$, is given by

$$\mathcal{H}X(z) := \tilde{X}(z) := \sum_{\alpha} c_\alpha z^\alpha \in \mathbb{C}. \tag{4.3}$$

We use multi-index notation here for z, i.e., $z = (z_1, z_2, \ldots) \in \mathbb{C}^N$ and $z^\alpha = z^{\alpha_1} z^{\alpha_2} \ldots$.

One of the most important results we use regularly in this chapter is the transformation of Wick products:

$$\mathcal{H}(X \diamond Y)(z) = \mathcal{H}(X)(z) \cdot \mathcal{H}(Y)(z) = \tilde{X}\tilde{Y}, \tag{4.4}$$

for all $z = (z_1, z_2, \ldots) \in \mathbb{C}^N$.

Another useful result we use is given by Lemma 2.8.4 in Holden et al. (2010), and it can be stated without any rigor, if

$$\frac{\mathrm{d}\tilde{X}(x,t)}{\mathrm{d}t} = \tilde{F}(t; z),$$

then

$$\frac{\mathrm{d}X(t,\omega)}{\mathrm{d}t} = F(t; \omega). \tag{4.5}$$

The expectation of $X(t, \omega)$, $E[X]$, is associated with the following properties:

$$E[X \diamond Y] = E[X] \cdot E[Y]; \quad X, Y \in (S)_{-1}, \text{ and} \tag{4.6}$$

$$E\left[\exp^{\diamond}[X]\right] = \exp[E[X]]; \quad X \in (S)_{-1}. \tag{4.7}$$

We illustrate the use of Hermite transform using a linear WSDE below.

Consider the following simple WSDE for anticipating $X(t)$,

$$\frac{\mathrm{d}X(t)}{\mathrm{d}t} = a + bX(t)\delta W(t), \quad X(0) = X_0, \tag{4.8}$$

where a, b, and X_0 are constants. Here the second term on the right-hand side leads to Skorohod integral, and using Eq. (4.1), we write,

$$\frac{\mathrm{d}X(t)}{\mathrm{d}t} = a + bX(t) \diamond w, \quad X(0) = X_0, \tag{4.9}$$

where w is one dimensional white noise, and X_0 is a constant.

Applying the Hermite transform to Eq. (3.8), we obtain,

$$\frac{\mathrm{d}\tilde{X}(t;z)}{\mathrm{d}t} = a + b\tilde{X}(t;z)\tilde{w}(t;z), \quad \tilde{X}(0) = X_0. \tag{4.10}$$

Here we see two properties of the Hermite transform: constants are invariant under the transformation and the white noise is transformed into the complex space. Now we can use the ordinary calculus to solve Eq. (4.10) analytically. The solution is given by with the simplification of notation,

$$\tilde{X}(t) = X_0 \exp\left[\int_0^t \left(b\tilde{X}(t)\tilde{w}(t)\right)\mathrm{d}u\right] + \int_0^t a \exp\left[\int_s^t b\tilde{w}(t)\mathrm{d}u\right]\mathrm{d}s. \tag{4.11}$$

Now we use the above-mentioned lemma (Eq. 4.5) to inverse transform Eq. (4.11):

$$X(t,\omega) = X_0 \exp^{\diamond}\left[\int_0^t (bX \diamond w(t,\omega))\mathrm{d}u\right] + \int_0^t a\exp^{\diamond}\left[\int_s^t (b\,w(t,\omega))\mathrm{d}u\right]\mathrm{d}s. \tag{4.12}$$

This can be simplified to,

$$\begin{aligned} X(t,\omega) &= X_0 \exp^{\diamond}\left[\int_0^t (bX \diamond \mathrm{d}W(t,\omega)\right] + \int_0^t a\ \exp^{\diamond}\left[\int_s^t (b\,\mathrm{d}W(t,\omega))\right]\mathrm{d}s, \\ &= X_0 \exp^{\diamond}\left[\int_0^t bX \diamond \mathrm{d}W\right] + \int_0^t a\ \exp^{\diamond}\left[\int_s^t b\,\mathrm{d}W\right]\mathrm{d}s, \\ &= X_0 \exp^{\diamond}\left[\int_0^t bX \diamond \mathrm{d}W\right] + \int_0^t a\ \exp^{\diamond}[b(W(t) - W(s))]\mathrm{d}s. \end{aligned} \tag{4.13}$$

As seen in Eq. (4.13), the initial value of X affects the first term of rhs, not the second term, and both a and b modify the second term of rhs.

If X is not anticipating, then we can write an Itô differential equation,

$$\mathrm{d}X(t,\omega) = a\mathrm{d}t + b\mathrm{d}W(t,\omega), \quad X(0) = X_0,$$

for the evolution of X. Then the analytical solution is,

$$X = X_0 + at + bW.$$

The solution of the WSDE and the solution of Itô SDE is characteristically different.

Following Holden et al. (2010), consider more general linear 1-dimensional Wick type Skorohod SDE as given in Eq. (3.1.1) in Holden et al. (2010),

$$\begin{aligned} \mathrm{d}X(t) &= g(t,\omega)\mathrm{d}t + r(t,\omega) \diamond X(t)\mathrm{d}t \\ &\quad + \sum_{i=1}^{m} \alpha_i(t,\omega) \diamond X(t)\delta W_i(t), \quad X(0,\omega) = X_0(\omega), \end{aligned} \tag{4.14}$$

where $g(t,\omega)$, $r(t,\omega)$, and $\alpha_i(t,\omega)$ are stochastic functions and they may be anticipating functions as well. Here the Wiener process is m-dimensional ($i = 1, \ldots, m$). It is assumed that all the stochastic functions and $X(t)$ are $(S)_{-1}$ valued.

Equation (4.13) can be written using the property of Skorohod integral (Eq. 4.1),

$$\frac{\mathrm{d}X(t)}{\mathrm{d}t} = g(t,\omega) + r(t,\omega) \diamond X(t)$$
$$+ \sum_{i=1}^{m} \alpha_i(t,\omega) \diamond X(t) \diamond w_i(t), \quad X(0,\omega) = X_0(\omega), \tag{4.15}$$

where $w(t) = (w_1(t), w_2(t), \ldots, w_m(t))$ is m-dimensional white noise.

Then Theorem 3.1.1 in Holden et al. (2010) gives the solution to Eq. (4.15) which is unique in $(S)_{-1}$:

$$X(t,\omega) = X_0(\omega) \diamond \exp^{\diamond}\left[\int_0^t \left(r(u,\omega) + \sum_{i=1}^{m} \alpha_i(u,\omega) \diamond w_i(u)\mathrm{d}u\right)\right]$$
$$+ \int_0^t \exp^{\diamond}\left[\int_s^t \left(r(u,\omega) + \sum_{i=1}^{m} \alpha_i(u,\omega) \diamond w_i(u)\mathrm{d}u\right)\right] \diamond g(s,\omega)\mathrm{d}s. \tag{4.16}$$

The expectation of $X(t,\omega)$ is given by,

$$E[X(t)] = E[X_0] \exp\left[\int_0^t E[r(u)\mathrm{d}u]\right]$$
$$+ \int_0^t \exp\left[\int_s^t E[r(u)\mathrm{d}u]\right] E\big[g(s)\big]\mathrm{d}s. \tag{4.17}$$

This theorem can be proved by using $\mathcal{H}$-transform. (See Holden et al. (2010).)

It is interesting to note that Eq. (4.15) has the form of a regular stochastic differential equation consisting of stochastic variables and coefficients. The noise terms on rhs of Eq. (4.15) are multiplicative with respect to $X(t)$, which is a realistic model in many physical situations. α_is may be modulating the externally induced fluctuations on the system, and g and r may depend on the intrinsic fluctuations. It may be possible that g, r, and α_is are influenced by both internal and external fluctuations. Therefore Eq. (4.15) could be a general model which may have wide ranging similarities to many real-world problems.

If we define

$$\psi(u,\omega) := r(u,\omega) + \sum_{i=1}^{m} \alpha_i(u,\omega) \diamond w_i(u),$$

then Eq. (4.16) can be written as,

$$X(t,\omega) = X_0(\omega) \diamond \exp^{\diamond}\left[\int_0^t \psi(u,\omega)\mathrm{d}u\right] + \int_0^t \exp^{\diamond}\left[\int_s^t \psi(u,\omega)\mathrm{d}u\right] \diamond g(s,\omega)\mathrm{d}s. \tag{4.18}$$

The Wick exponential terms in Eq. (4.16) are positive, and if $X_0(\omega)$ and $g(s,\omega)$ are always positive processes then $X(t,\omega)$ evolves in $\mathbb{R}^+$. If $r(u,\omega)$ and $g(s,\omega)$ can be expressed as Itô differentials, the expectation of $X(t,\omega)$ is a deterministic function according to Eq. (4.17). If we can assume X_0, g, r, and α_is are real valued constants, using Eq. (4.18), Eq. (4.17) can be further simplified.

$$E[X(t)] = \left(X_0 + \frac{g}{r}\right)\exp[rt] - \frac{g}{r}. \tag{4.19}$$

From Eq. (4.19), the effects of X_0, g, and r on the expectation are such that X_0 and $\frac{g}{r}$ would determine the behavior of $E[X(t)]$ for a given nonzero value of r.

It should be mentioned that if X_0 and α_is are real numbers, and g and r are deterministic functions, then Eq. (4.16) can be written without Wick products, and Eq. (4.18) is a deterministic function of t.

4.3 Models of Positive White Noise

As we have discussed in Chap. 2, given the d-parameter Wiener process, we can define the Wiener–Itô integral of a function ϕ $\left(\in L^2\left(R^d\right)\right)$ as,

$$\int \phi(x)\mathrm{d}W(x,\omega) := \langle \omega, \phi \rangle, \tag{4.20}$$

integrating over R^d, $\omega \in S'\left(R^d\right)$, where $\langle \omega, \phi \rangle$ is the action ω on ϕ, also denoted as $w(\phi)$ (Holden et al., 2010).

We can construct a smooth white noise process $w_\phi(x,\omega)$ that can be thought of as a white noise process smoothen by a continuous function, for example, a test function with bounds, which is normally distributed with mean 0 and variance, $\|\phi\|^2$. $w_\phi(x,\omega)$ could be a useful model in applications.

Holden et al. (2010) show that (see Lemma 2.6.16),

$$\exp^{\diamond}\left[w_\phi(x,\omega)\right] = \exp\left[w_\phi(x,\omega) - \frac{1}{2}\|\phi\|^2\right], \tag{4.21}$$

and both sides are positive.

We can model a time dependent coefficient, k, with internal noise as,

$$k(t,\omega) = k_0 \exp^{\diamond}\big[a\, w_\phi(x,\omega)\big], \tag{4.22}$$

where k_0 is a constant and a is an amplitude factor of the smooth white noise.

In many physical and chemical systems, irregularities associated with a coefficient are positive, fluctuating around a mean value. For example, k_0 could be a mean reaction rate of a chemical reaction, and the dynamic reaction rate could be modeled as Eq. (4.22). This can be seen by taking the mean of Eq. (4.22). If a is a constant,

$$E[k(t,\omega)] = k_0.$$

According to the Wick exponential expansion,

$$\exp^{\diamond}\big[aw_\phi\big] = \exp\left[aw_\phi - \frac{1}{2}\|\phi\|^2\right] = \exp\left[-\frac{1}{2}\|\phi\|^2\right]\exp\big[aw_\phi\big]. \tag{4.23}$$

From Eq. (4.21), we see that the noise term would be mediated by ϕ and, for example, if $\phi = 1$, then $\|\phi\|^2 = t$; hence we see a decrease in the noise term over time, which may not be a suitable model. However, if we discretize t, and the beginning of ith interval is t_i, a constant, then we can approximate,

$$k(t_i,\omega) \approx k_0 \exp[a\, w(t_i,\omega)] \tag{4.24}$$

which gives us a stochastic model for coefficients that can be easily incorporated in computations.

References

Di Nunno, G., Øksendal, B., & Proske, F. (2019). *Malliavin calculus for Lévy processes with applications to finance*. Springer-Verlag.

Holden, H., et al. (2010). *Stochastic partial differential equations*. Springer New York. https://doi.org/10.1007/978-0-387-89488-1

Lindstrøm, T., Øksendal, B., & Ubøe, J. (1991). Stochastic differential equations involving positive noise. In M. Barlow & N. Bingham (Eds.), *Stochastic analysis* (pp. 261–303). Cambridge University Press.

Chapter 5
Formulation of Numerical Solutions for Wick Type Stochastic Differential Equations (WSDEs)

5.1 Introduction

As we have seen in Chaps. 2, 3, and 4, one of the drawbacks of using WSDEs in modeling is difficulty in computing analytical solutions. In this chapter, we introduce the formulation of a numerical scheme that could be used in some practical situations. We assume that a stochastic variable of interest can be expressed as a time dependent mean plus a stochastic component, and this assumption is good enough for a variety of situations we encounter in chemical transformations. As we have seen in Chap. 1, we can consider concentrations of chemical species as having mean components plus the fluctuating components due to intrinsic and external noise in the system under consideration. Then we can include the stochastic variables (concentrations) in the formulation of the model. This assumption is supported by the fact that in situations where the chemically reacting systems do not have any noise, there are always nonnegative and deterministic concentrations of chemical species. We can test this using WSDEs and experimental data of average values of concentrations.

In situations where $X(t, \omega) \in L^2(P)$ is $\mathcal{F}_t$—measurable then there is a unique process,

$$\psi(t, \omega), 0 \leq t \leq T, \text{ and}$$

$$X(t, \omega) = E[X(t, \omega)] + \int_0^t \psi(t, \omega) dW(t, \omega). \tag{5.1}$$

$\psi(t, \omega)$ is $\mathbb{F}$—adapted, and Eq. (5.1) is the Itô representation theorem that would express $X(t, \omega)$ as an Itô process but the Itô representation theorem does not provide an explicit expression for $\psi(t, \omega)$. But by using Clark-Ocone formula (Nunno et al., 2019), we can interpret $\psi(t, \omega)$ using the Malliavin derivative:

D. Kulasiri, *Stochastic Differential Equations for Chemical Transformations in White Noise Probability Space*, https://doi.org/10.1007/978-981-97-9392-1_5

$$\psi(t,\omega) = E[(D_t X(t,\omega)|\mathcal{F}_t)], 0 \le t \le T, \tag{5.2}$$

where $D_t X(t,\omega)$ is the Malliavin derivative of $X(t,\omega)$. The Malliavin derivative can be thought of as a stochastic gradient in the direction of ω (Nunno et al., 2019).

Even though the Clark-Ocone formula and the Itô representation theorem have similar mathematical expressions, they are conceptually very different results. The Clark-Ocone formula connects Sobolev space differential calculus and the Itô calculus. We make use of this connection by expressing $X(t,\omega)$ in terms of the expected value and the integral given in Eq. (5.1) and Eq. (5.2).

5.2 An Approximate Numerical Solution for WSDEs

Let us revisit Eq. (4.15) in Chap. 4 and consider X_0, g, r are constants to simplify the development first (Holden et al. 2010). $W(t,\omega)$ is the one-dimensional Wiener process, $w(t,\omega)$ is the one-dimensional white noise process and $\alpha_i = \alpha$, a constant. Then we can write Eq. (4.15) as,

$$\frac{dX(t,\omega)}{dt} = g + r\Diamond X(t,\omega) + \alpha\Diamond X(t,\omega)\Diamond w(t,\omega), X(0,\omega) = X_0. \tag{5.3}$$

Given that the coefficients are constants,

$$\frac{dX(t,\omega)}{dt} = g + rX(t,\omega) + \alpha X(t,\omega)\Diamond w(t,\omega). \tag{5.4}$$

Let us assume that the condition of the process is such that we can express,

$$X(t,\omega) = E[X(t,\omega)] + \int_0^t \psi(t,\omega)dW(t,\omega), \tag{5.5}$$

and ψ is the Malliavin derivative of X. Let us denote

$$E[X(t,\omega)] = f(t), \tag{5.6}$$

and

$$\mathcal{S}(t,\omega) := \int_0^t \psi(t,\omega)dW(t,\omega). \tag{5.7}$$

Then,

$$X(t,\omega) = f(t) + \mathcal{S}(t,\omega), \tag{5.8}$$

and $f(t)$ is a deterministic function. Substituting, Eq. (5.8) to Eq. (5.4),

$$\begin{aligned}\frac{df(t)+\mathcal{S}(t,\omega)}{dt} &= g + r(f(t)+\mathcal{S}(t,\omega)) \\ &\quad + \alpha\big[f(t)+\mathcal{S}(t,\omega)\big]\diamond w(t,\omega).\end{aligned} \tag{5.9}$$

This leads to,

$$\begin{aligned}\frac{df(t)}{dt} + \frac{d\mathcal{S}(t,\omega)}{dt} &= g + rf(t) + r\mathcal{S}(t,\omega) + \alpha f(t)w(t,\omega) \\ &\quad + \alpha\mathcal{S}(t,\omega)\diamond w\end{aligned} \tag{5.10}$$

From Eq. (5.10), we see that,

$$\frac{df(t)}{dt} = g + rf(t), \tag{5.11}$$

leading to,

$$\frac{d\mathcal{S}(t,\omega)}{dt} = r\mathcal{S}(t,\omega) + \alpha f(t)w(t,\omega) + \alpha\mathcal{S}(t,\omega)\diamond w(t,\omega). \tag{5.12}$$

We can discretize Eqs. (5.11) and (5.12) if $\Delta t = t_{i+1} - t_i$, an infinitesimal equidistant time interval within [0,T]. Equation (5.11) can be solved analytically or numerically; and Eq. (5.12) can be discretized using simple finite differences after multiplying by dt:

$$\begin{aligned}\mathcal{S}(t_{i+1}) - \mathcal{S}(t_i) &\cong r\mathcal{S}(t_i,\omega)dt + \alpha f(t_i)dW(t_i,\omega) \\ &\quad + \alpha\mathcal{S}(t_i,\omega)\diamond dW(t_i,\omega).\end{aligned} \tag{5.13}$$

Note that we use the increments of the Wiener process and rhs is evaluated at t_i. We use Δt of the order of 10^{-4} to10^{-5} in computations. According to Eq. (3.9),

$$\begin{aligned}\mathcal{S}(t_i,\omega)\diamond dW(t_i,\omega) &\cong \mathcal{S}(t_i,\omega)\diamond(W(t_{i+1},\omega) - W(t_i)) \\ &= \mathcal{S}(t_i,\omega)(W(t_{i+1},\omega) - W(t_i,\omega)) \\ &= \mathcal{S}(t_i,\omega)dW(t_i,\omega).\end{aligned} \tag{5.14}$$

Hence,

$$\begin{aligned}\mathcal{S}(t_{i+1}) &\cong \mathcal{S}(t_i) + r\mathcal{S}(t_i,\omega)dt + \alpha f(t_i)dW(t_i,\omega) \\ &\quad + \alpha\mathcal{S}(t_i,\omega)dW(t_i,\omega),\ \mathcal{S}(0) = 0\end{aligned} \tag{5.15}$$

and

$$X(t_{i+1}) \cong f(t_{i+1}) + \mathcal{S}(t_{i+1}, \omega) \tag{5.16}$$

Based on this finite difference scheme, we develop numerical schemes using Mathematica® and the code is given in Appendix 5.1.

5.3 Computations

We evaluate the numerical solutions of WSDE (Eq. (5.3)) for $\Delta t = 0.0001$; N $=$ 10000; $g = 0.4$; $r = 0.1$; $\alpha = 0.1$; $X_0 = 1$. Here N is the total number of intervals, hence the final time $T = 1.0$. Figure 5.1 shows the Wiener realization used for the computation of the solution X.

The approximate solution of Eq. (5.3) stemming from the Wiener realization in Fig. 5.1 is given in Fig. 5.2.

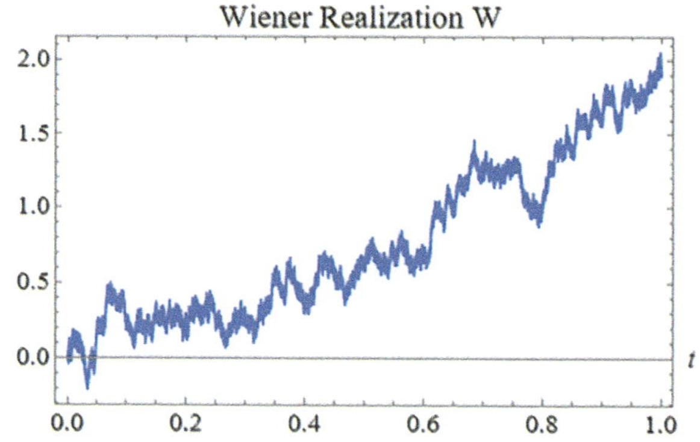

Fig. 5.1 The realization of the Wiener process that is used for the subsequent computations

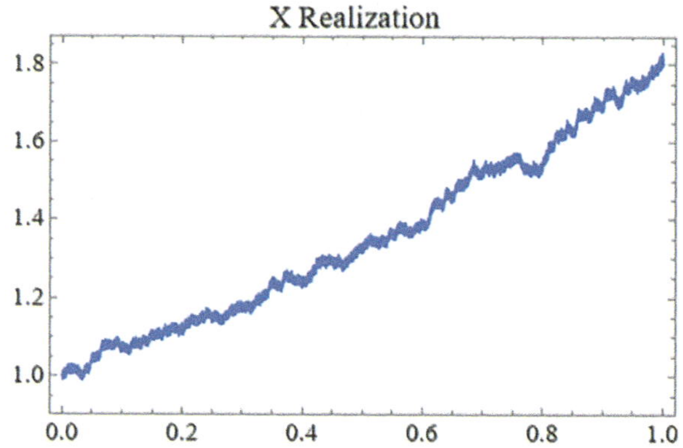

Fig. 5.2 The approximate solution of WSDE for the given Wiener process and for $g = 0.4$; $r = 0.1$; $\alpha = 0.1$; $X_0 = 1$

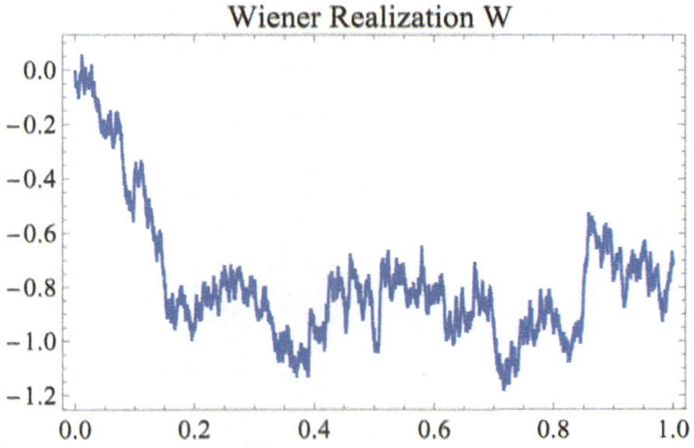

Fig. 5.3 Another realization of the Wiener process

It is seen that for $\alpha = 0.1$, X follows the trend of the Wiener process. When we compute the Wiener process again, the realization changes to the one shown in Fig. 5.3, which shows a dramatic change in the process.

When we compute the approximate X, we see that the X realization (Fig. 5.4) has significantly lower values at each instance of time as compared to Fig. 5.2.

The stochastic nature of the solution is obviously determined by the Wiener process for the same α, and this is very similar to the experiments in complex systems influenced by fluctuations, hence we need to take an adequate number of realizations to calculate to understand the behavior of X. For this particular realization, we overlay the expected value of X with the realization of X (Fig. 5.5). We compute the expected value based on the deterministic solution of the WSDE, but we can also compute the expected value using a large number of realizations and taking the mean at each time instance, which we compute later.

It is noted in Fig. 5.5 that the expected value of X is higher than the realization of X. In the second part of the program in Appendix 5.1 computes 30 realizations of the Wiener process and the corresponding realizations of X without changing the

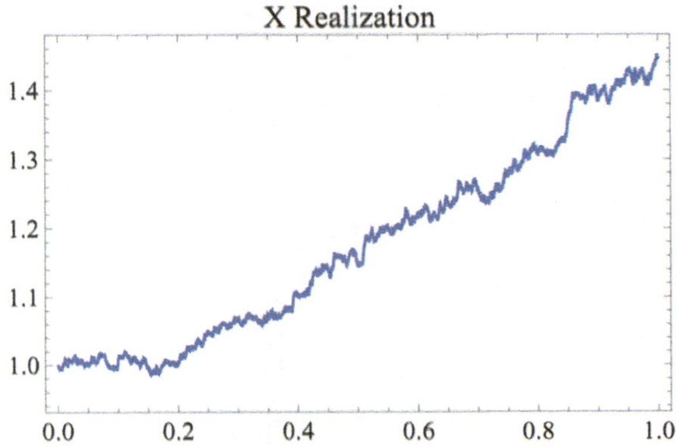

Fig. 5.4 The realization of X corresponding to the Wiener realization shown in Fig. 5.3 for the same values of coefficients

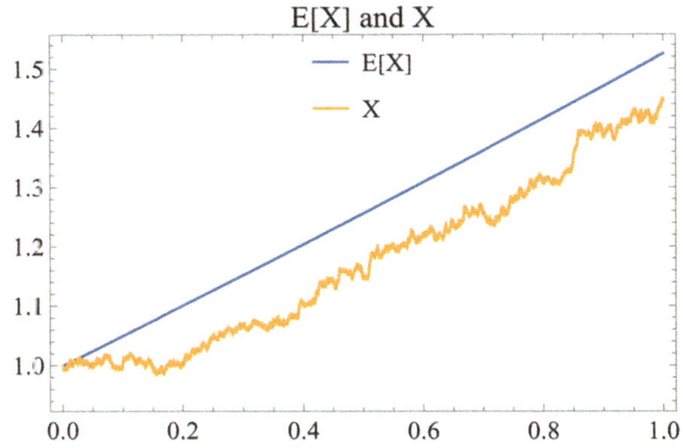

Fig. 5.5 The expected values of X and a realization of X

values of coefficients. Figure 5.6 shows the 30 realizations of X computed using 30 realizations of the Wiener process.

The mean of X realizations is given in Fig. 5.7.

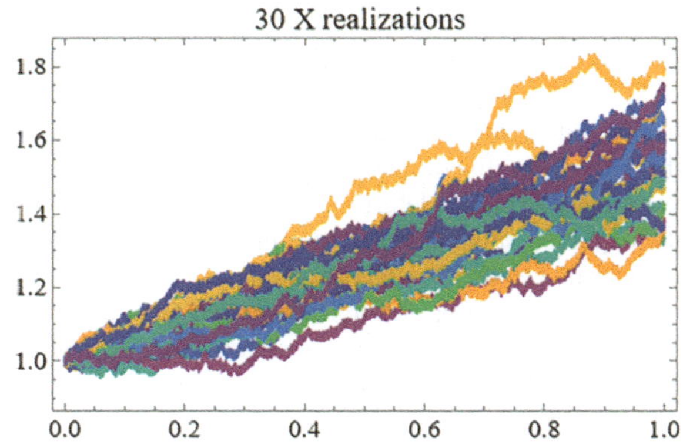

Fig. 5.6 30 realizations of X for the same coefficients as before

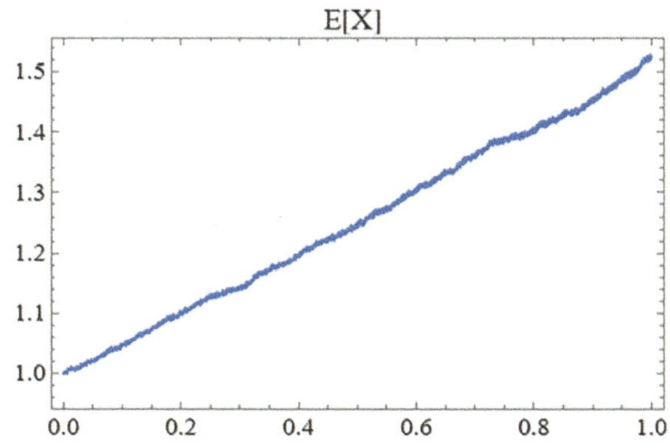

Fig. 5.7 The mean of 30 realizations shown in Fig. 5.6

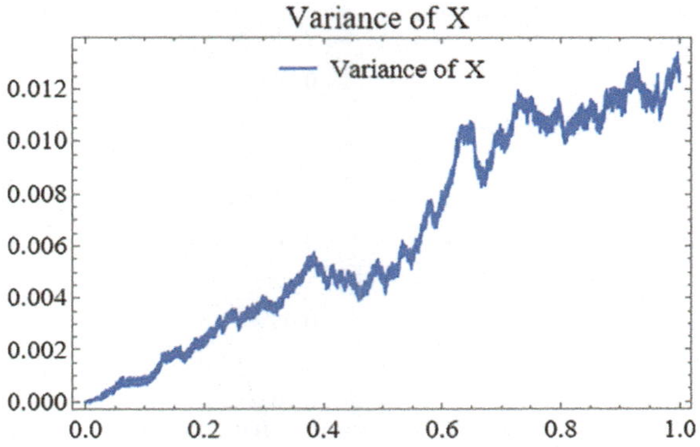

Fig. 5.8 Variance of X of the 30 realizations shown in Fig. 5.6

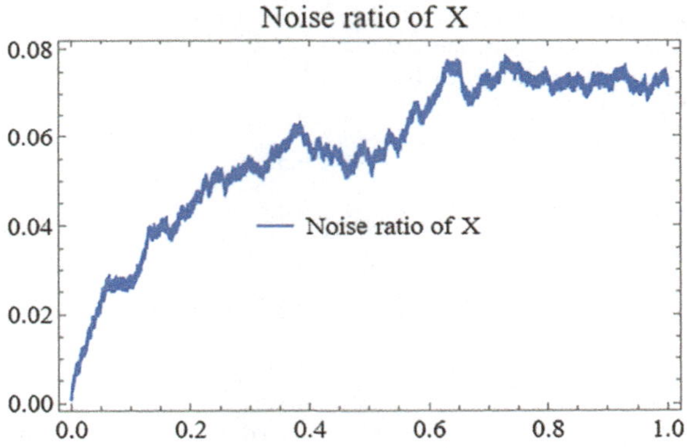

Fig. 5.9 Noise ratio of X based on 30 realizations for $\alpha = 0.1$

We see that some of the realizations overshoot the mean and others undershoot the mean. We also compute the variance of the 30 realizations (Fig. 5.8), and then we compute the noise ratio as defined by the square root of variance divided by the mean (Fig. 5.9).

Noise ratio is a good measure of the influence of fluctuations for a given set of coefficients, and we can change the coefficients to gauge the effects on the using the noise ratio. In this case, the noise ratio reaches a stable value around 0.07 for larger values of time, which is an interesting observation that needs to be explored further.

The third part of the program computes the noise ratios for two different α values (0.2 and 0.5) using the same set of Wiener increments of 30 realizations. Figures 5.10 and 5.11 show the noise ratios of X when $\alpha = 0.2$ and $\alpha = 0.5$, respectively.

Noise ratios increase significantly with α and the choice of which determines the trajectories of X but the trends are not influenced by the Wiener processes used. The time derivatives of noise ratio tend to decrease over a long time.

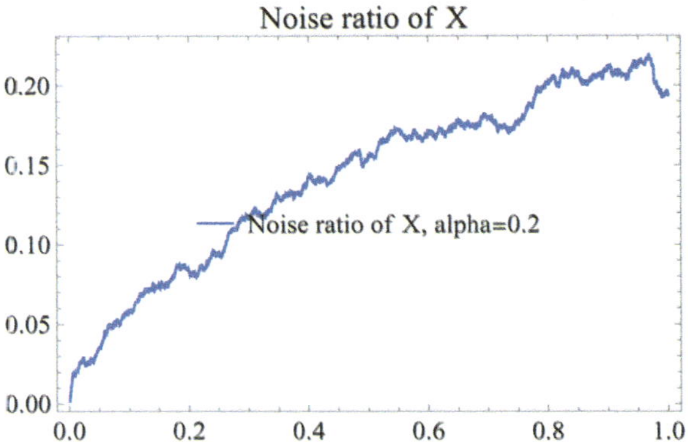

Fig. 5.10 Noise ratio of X when $\alpha = 0.2$

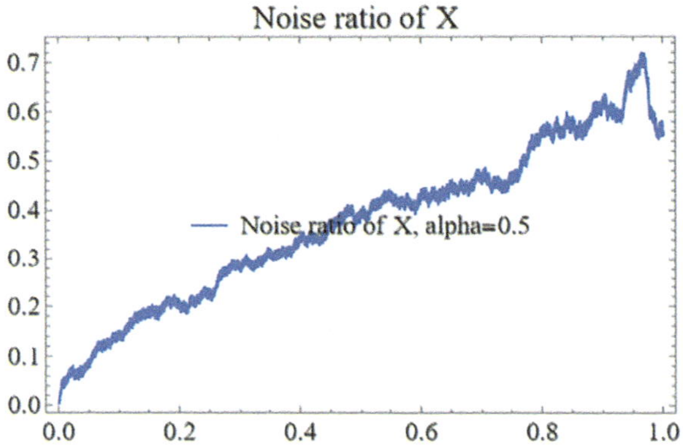

Fig. 5.11 Noise ratio of X when $\alpha = 0.5$ using the same set of Wiener processes as in Fig. 5.10

Appendix 5.1: Mathematica® Code Without Outputs

(*We start with the SDE dX/dt = g(t,*) + r(t,*) ◇ X(t,*) + α(t,*)◇ X(t,*) ◇ w, where w is the white noise. Here g, r, and α may be stochastic functions but we assume them to be constants or deterministic functions; then the equation becomes dX/dt = g + r X + α X ◇ w, only the last term has a Wick product*).

(*Computation of increments of Wiener process*)

$$\Delta t = 0.0001;\ \mathrm{N} = 10000;\ g = 0.4;\ r = 0.1;\ \mathrm{alpha} = 0.1;\ \mathrm{X0} = 1.;$$

$$\Delta\mathrm{WList} = \mathrm{Table}[N[\mathrm{Random}[\mathrm{NormalDistribution}\ 0., \mathrm{Sqrt}[\Delta \mathrm{t}]]], 32], \{\mathrm{N}\}];$$

(*Create *a*list with time in (*y*, *t*)format*)

```
ΔWtList = Table[{(i − 1)Δt, ΔWList[[i]]}, {i, 1, N}];
```

(*Plotting*)

```
ListLinePlot[ΔWtList, Frame → True, Axes → True,
Joined → True,
PlotLabel →" Wiener Increments ΔW",
AxesLabel → {HoldForm[t], HoldForm[ΔW]},
ColorFunction → Automatic,
LabelStyle− > {FontFamily
        →" Times New Roman", 14, GrayLevel[0]}]
```

(*Computation of Wiener process*)

```
WP = Accumulate[ΔWList];
WPtList = Table[{(i − 1)Δt, WP[[i]]}, {i, 1, N}];
```

(*Plotting*)

```
ListLinePlot[WPtList, Frame → True, Axes → True,
Joined → True, PlotLabel →" Wiener Realization W",
AxesLabel → {HoldForm[t]}, ColorFunction → Automatic,
LabelStyle− > {FontFamily
        →" Times New Roman", 14, GrayLevel[0]}]
```

(*Calculate the deterministic function *)

```
fList = Table[−(g/r) + (X0 + (g/r))Exp[rΔt(i − 1)],
      {i, 1, N + 1}];
```

(*Now we calculate the stochastic component of the equation S*)

```
SList = Table[0., {i, 1, N + 1}];

Do[SList[[i + 1]]
      = SList[[i]] + rSList[[i]]Δt + alpha(fList[[i]]
      +SList[[i]])ΔWList[[i]], {i, 1, N}];

XList = SList + fList;
```

```
XtList = Table[{(i − 1)Δt, XList[[i]]}, {i, 1, N}];
```

(*Plotting*)

```
ListLinePlot[XtList, Frame → True, Axes → True,
Joined → True,
PlotLabel → "X Realization", AxesLabel → {HoldForm[t]},
ColorFunction → Automatic,
LabelStyle− > {FrontFamily
        → "Times New Roman", 14, GrayLevel[0]}]
```

(*Calculate the expected value*)

```
XeList = Table[−(g/r) + (X0 + g/r)(Exp[rΔt(j − 1)]),
        {j, 1, N + 1}];

XetList = Table[{(i − 1)Δt, XeList[[i]]}, {i, 1, N}];
```

(*Plotting*)

```
ListLinePlot[{XetList, XtList}, Frame → True, Axes → True,
Joined → True, PlotLabel → "E[X] and X",
AxesLabel → {HoldForm[t]}, ColorFunction → Automatic,
PlotLegends → Placed[{E[X], X}, {Center, Top}],
LabelStyle− > {FontFamily
        → "Times New Roman", 14, GrayLevel[0]}]
```

(* Second part: Computing 30 realizations of X*)

```
ΔWList = Table[N [Random [NormalDistribution
    [0., Sqrt[Δt]]], 32], {N}, {30}];
fList = Table[−(g/r) + (X0 + (g/r))
    Exp[rΔt(i − 1)], {i, 1, N + 1}, {j, 1, 30}];
ftList = Table[{(i − 1)Δt, fList[[i]]}, {i, 1, N + 1}, {j, 1, 30}];
SList = Table[0., {i, 1, N + 1}, {j, 1, 30}];
Do[SList[[i + 1, j]]
    = SList[[i, j]] + rSList[[i, j]]Δt + alpha(fList[[i, j]]
    +SList[[i, j]]ΔWList[[i, j]], {i, 1, N}, {j, 1, 30}];
```

```
XList = fList + SList;
XListT = Transpose[XList];
XtListT = Table[{(i − 1)Δt, XListT[[j, i]]}, {j, 1, 30}, {i, 1, N + 1}];
```

(*Plotting*)

```
ListLinePlot[XtListT, Frame → True, Axes → True,
Joined → True, PlotLabel → 30 X realizations,
AxesLabel → {HoldForm[t]}, ColorFunction → Automatic,
LabelStyle− > {FontFamily
        → "Times New Roman", 14, GrayLevel[0]}]

ListLinePlot[Mean[XtListT], Frame → True, Axes → True,
Joined → True, PlotLabel → "E[X]",
AxesLabel → {HoldForm[t], HoldForm[X]},
ColorFunction → Automatic,
LabelStyle− > {FontFamily
        → "Times New Roman", 14, GrayLevel[0]}]
```

(*Calculating Noise ratio *)

```
meanXListT = Mean[XtListT];
varXListT = Variance[XtListT];
varXtList = Table[{Δt(i − 1), varXListT[[i, 2]]}, {i, N + 1}];
meanXtList = Table[{Δt(i − 1), meanXListT[[i, 2]]}, {i, N + 1}];
```

(*Plotting*)

```
ListLinePlot[varXtList, Frame → True, Axes → True,
Joined → True, PlotLabel → "Variance of X",
AxesLabel → {HoldForm[t], HoldForm[Var[X]]},
ColorFunction → Automatic,
PlotLegends → Placed[{Variance of X}, {Center, Top}],
LabelStyle− > {FontFamily
        → "Times New Roman", 14, GrayLevel[0]}]
```

(*Compute noise ratio*)

```
noiseRatio =
```

```
Table[{Δt(i
 −1), Sqrt[varXtList][[i, 2]]/meanXtList[[i, 2]]},
 {i, 2, N + 1}];
```

(*Plotting*)

```
ListLinePlot[{noiseRatio}, Frame → True, Axes → True,
Joined → True, plotLabel → Noise ratio of X,
AxesLabel → {HoldForm[t]}, ColorFunction → Automatic,
PlotRange → All,
plotLegends → Placed Noise ratio of X, {Center, Center},
LabelStyle− > {FontFamily
        → "Times New Roman", 14, GrayLevel[0]}]
```

(*Third Part: Noise ratios for two different alpha values. The same set of 30 Wiener increments are used in both cases*)

```
Δt = 0.0001; N = 10000; g = 0.4; r = 0.1; alpha = 0.5; X0 = 1.;
```

```
ΔWList = Table[N [Random [NormalDistribution
      [0., Sqrt[Δt]]], 32, {N}, {30}];
fList = Table[−(g/r) + (X0 + (g/r))Exp[rΔt(i − 1)],
      {i, 1, N + 1}, {j, 1, 30}];
ftList = Table[{(i − 1)Δt, fList[[i]]}, {i, 1, N + 1}, {j, 1, 30}];
SList = Table[0., {i, 1, N + 1}, {j, 1, 30}];
Do[SList[[i + 1, j]]
        = SList[[i, j]] + rSList[[i, j]]Δt
        + alpha(fList[[i, j]]
        +SList[[i, j]])ΔWList[[i, j]], {i, 1, N}, {j, 1, 30}];
XList = fList + SList;
XListT = Transpose[XList];
XtListT = Table[{(i − 1)Δt, XListT[[j, i]]}, {j, 1, 30}, {i, 1, N + 1}];
```

(*Plotting*)

```
ListLinePlot[XtListT, Frame → True, Axes → True,
Joined → True, PlotLabel → 30 X realizations for alpha = 0.2,
```

```
AxesLabel → {HoldForm[t]}, ColorFunction → Automatic,
LabelStyle− > {FontFamily
        → "Times New Roman", 14, GrayLevel[0]}]
```

(*Calculating Noise ratio*)

```
meanXListT = Mean[XtListT];
varXListT = Variance[XtListT];
varXtList = Table[{Δt(i − 1), varXListT[[i, 2]]}, {i, N + 1}];
meanXtList = Table[{Δt(i − 1), meanXListT[[i, 2]]}, {i, N + 1}];
noiseRatio =
Table[{Δt(i
   −1), Sqrt[varXtList][[i, 2]]/meanXtList[[i, 2]]}, {i, 2, N + 1}];
```

(*Plotting*)

```
ListLinePlot[{noise Ratio}, Frame → True, Axes → True,
Joined → True, PlotLabel → Noise ratio of X,
AxesLabel → {HoldForm[t]}, ColorFunction → Automatic,
PlotRange → All,
PlotLegends → PlaceNoise ratio of X, alpha = 0.2, {Center, Center},
LabelStyle → {FontFamily
        → "Times New Roman", 14, GrayLevel[0]
```

(*Similar computations can be repeated in a module for various alpha values, but this is not shown here*).

References

Di Nunno, G., Øksendal, B., & Proske, F. (2019). *Malliavin calculus for Lévy processes with applications to finance*. Springer.

Holden, H., et al. (2010). *Stochastic partial differential equations*. Springer New York. Available at: https://doi.org/10.1007/978-0-387-89488-1.

Kondepudi, D., & Prigogine, I. (2014) *Modern thermodynamics: from heat engines to dissipative structures* 2nd edn. Available at: https://doi.org/10.1002/9781118698723.

Chapter 6
An Investigation of Deterministic Chemical Reactions and Their Stability

6.1 Introduction

In Chap. 1, we discuss very briefly the role of thermodynamics in chemical reactions. The well-established equilibrium thermodynamics renders a solid understanding of the role of chemical transformations including phase transitions in systems. Chemical transformations or reactions must satisfy the mass and energy conservation laws, and also the second law of thermodynamics. The second law of thermodynamics introduces entropy as a measure of disorder, i.e., higher the disorder higher the entropy of the system, but often we see this to be true in closed systems or isolated systems. In these systems, the entropy reaches a maximum and the system establishes itself in a stationary state. A perturbation from the stationary state would restore the entropy to the maximum, implying that the entropy change due to perturbation should be negative. Based on the Gibb's equation on internal energy as a function of chemical potential, De Donda (Kondepudi & Prigogine, 2014) introduced the concept of affinity as a "driving force" of a chemical reaction. The idea of affinity has the origin in Newtonian mechanics where the concept of force is introduced. Similarly, the affinity having the origin in the thermodynamic force dependent on chemical potential and the movement of molecules or flux is based on the idea of extent of reaction. The extent of reaction is defined to quantify the moles changes in each species involved in a reaction. Given this definition the derivative of extent of reaction is linearly related to thermodynamics flux if the system volume is a constant, and the time derivative of the entropy internal to the system is expressed in the form of the thermodynamic force multiplied by the thermodynamic flux. This formulation of internal rate of change of entropy has far reaching implications. In an isolated system, entropy reaches a maximum at stable equilibrium point, and to this to happen either thermodynamic flux and/or force must be zero. This is true for isolated or closed systems, but open systems behave rather unpredictably in relation to stability. One important point to note however, the Gibb's free energy of a reaction and the affinity differ only by a negative sign, i.e., the absolute numerical

D. Kulasiri, *Stochastic Differential Equations for Chemical Transformations in White Noise Probability Space*, https://doi.org/10.1007/978-981-97-9392-1_6

values are equal. As discussed in Kondepudi and Prigogine (2014), conceptually however, affinity relates the irreversible chemical processes, whereas the Gibb's free energy is used in reversible reactions in equilibrium states. Hence affinity is a broader concept associated with irreversible chemical processes. We use this relationship to understand stability issues of open systems.

As we mentioned previously, the affinity can be calculated by chemical potentials of species participating in a reaction. According to the second law, affinity (force) multiplied by the rate of change of reaction (velocity of reaction or reaction rate) should be nonnegative. This allows us to decipher the net direction of an irreversible reaction: if affinity is positive, the net reaction proceeds to the right, and if the affinity is negative, the reaction proceeds to the left. At equilibrium, the affinity is zero where the entropy production is zero as well, but the entropy is a maximum.

We see how the velocity of reaction enters into the calculation of rate of entropy production. The velocity of reaction per unit volume is defined using the rate of change of the extent of reaction, which is the independent variable for that reaction as the concentrations of species can be derived according to mass conservation law. The extent of reaction or reaction velocity is not determined by the thermodynamics except for the fact that the temperature of the medium would affect the reaction rate according to the Arrhenius law. Hence, we express chemical reaction rate according to kinetic empirical laws such as mass action, and Micheal-Menten for the reactions where enzymes play catalytic roles. Hill function is used where chemical affinities are strong and related to proteins as in gene transcription and translation. In this book, we only use mass action kinetics because our emphasis is to develop mathematics related to stochastic chemical transformations. The methods developed can also be used in the situations where other kinetic laws are more appropriate. However, specific dynamics can be affected by the choice of kinetic laws, therefore care must be shown in this choice based on experimental evidence.

While the closed systems are quite useful in studying the constrained systems in laboratories, almost of all systems in nature are open to the mass and energy flows across the boundaries. Then we enter the realm of nonlinear thermodynamics where the systems behave near thermodynamic equilibrium. The linear nonequilibrium theory developed by Onsager is applicable near equilibrium where in the linearization of thermodynamic fluxes and forces is introduced. The intra- and inter-cellular processes in biology are highly organized to function reliably and efficiently but operate far from equilibrium. This means the entropy must decrease in these situations, but the entropy flows in and out of the boundaries behave such a way that the second law is not violated. But these situations are not simple as they sound, hence we need to look into them further.

In the linear regimes of nonlinear thermodynamics, reactions operate near the equilibrium, hence the stability of reactions can be restored upon perturbation. We could model the entropy production per unit volume of a reaction using similar formulation discussed before, i.e., using affinity and velocity of reaction. In these situations, Onsager's linear phenomenological laws can be incorporated to develop entropy production formulation with microscopic reversibility automatically included. (See Kondepudi and Prigogine (2014).)

In open systems, the self-organization happens through the establishment of NESSs, fluctuations, and oscillations. Often fluctuations induce the migration of NESSs to different states or instability or oscillations. These unpredictable behaviors do not adhere to any extremum principles and therefore are very much context driven. Hence, each situation should be analyzed according to the particular reaction scheme. The emergence of NESSs and associated oscillating concentrations around them is due to reactive and diffusive properties of the species in the system. We can also use stochastic approaches to understand the fluctuations driven systems in addition to dynamical system approaches.

We use two approaches to investigate the reactions: using the excess entropy production conditions and the dynamic stability theory of ODEs. First, we briefly summarize the excess entropy production conditions and discuss the stability conditions of ODEs in a simple setting. The purpose of this chapter is to summarize essential ideas so that we are familiar with them when we invoke them in later chapters. There are excellent references are available for further study; a very few of them are given in the references.

We mention in Chap. 1 following (Kondepudi & Prigogine, 2014), the entropy production P in a NESS is used to determine stability of NESS. Following Lyapunov stability theory, we see that the second variation of entropy $\left(\delta^2(S)\right)$ is a Lyapunov functional, and the change in entropy from the NESS (ΔS) can be written as,

$$\Delta S = \delta S + \frac{1}{2}\delta^2 S. \tag{6.1}$$

It can be shown that,

$$\frac{\mathrm{d}\left(\frac{\delta^2 S}{2}\right)}{\mathrm{d}x} = \int \sum_k \delta F_k \delta J_k \mathrm{d}V > 0 \tag{6.2}$$

for a NESS to be stable. Here the integral is over the system volume, k is the reaction number, δF_k is an infinitesimal change in the thermodynamic force of kth reaction and δJ_k is an infinitesimal change in the thermodynamic flow of kth reaction. If this inequality is violated, it indicates that the system may be unstable, but it is not a sufficient condition for instability. We can use this criterion to investigate stability of a NESS using appropriate ODEs. In deriving the relevant ODEs, we find the use of the extent of reaction is meaningful for a single reaction, but for a system of reactions, it is less cumbersome to use the concentration of species. We use both approaches to investigate the stability of NESSs in this chapter.

Stability investigation of a dynamic system using ODEs is a highly developed subject and there are many excellent references available. Here we introduce key concepts that we use later. An ODE can be expressed by

$$\frac{\mathrm{d}y}{\mathrm{d}t} = f(t, y, \lambda), \tag{6.3}$$

where t and y are independent and dependent variables, respectively, and λ is a parameter that is introduced in the derivation of the ODE. At a NESS or critical point,

$$\frac{\mathrm{d}y}{\mathrm{d}t} = 0, \text{ i.e., } f(t, y, \lambda) = 0. \tag{6.4}$$

The solution to this equation provides the stability values y_s at t_s, for a particular value of λ. As we change λ, y_s moves along a straight line in a plot y_s versus λ, i.e., NESS is still maintained but in a critical point of λ, y_s curve folks into two branches. The λ value at which the symmetry is broken is called the bifurcation point (λ_b), fluctuations may determine along with other conditions which branch the system would move into. When $\lambda < \lambda_b$ but very near to λ_b, we can write

$$y = y_s + \delta y, \tag{6.5}$$

where δy is an infinitesimal perturbation from y_s, then,

$$\frac{\mathrm{d}\delta y}{\mathrm{d}x} = f(t, y_s + \delta y, \lambda). \tag{6.6}$$

We put $t = 0$, when we introduce the perturbation and solve Eq. (6.6) for time evolution of perturbation. If NESS is stable δy should diminish toward zero, otherwise the system is not stable. We use entropy investigation and this simple perturbation analysis to investigate the chemical reactions given by Kondepudi and Prigogine (2014) to illustrate different behaviors of the systems.

6.2 Example 1

We follow closely Kondepudi and Prigogine (2014) in these examples, so that readers can follow them in more detail in their book. Our intention here is to show how entropy, perturbation analysis, and numerical solutions are used to understand the system behavior first in a closed system in which a reaction may reach thermodynamic equilibrium state. Then we use the same reaction in an open system in which the reaction may reach a NESS. We assume that the inflows and outflows are controlled in the system so that a NESS is plausible.

The first reaction we consider is the following:

$$A + B \rightleftharpoons C + D, \tag{6.7}$$

and k_f and k_r are forward and reverse reaction rate coefficients. We use mass action law for elementary reactions to write the reaction rates in forward and reverse directions:

$$R_{\mathrm{f}} = k_{\mathrm{f}}[A][B], \tag{6.8}$$

and

$$R_{\mathrm{r}} = k_{\mathrm{r}}[C][D]. \tag{6.9}$$

Because we are operating in a closed system, the extent of reaction is used to model the reaction dynamics; however, in an open system the extent of reaction is not appropriate as it is far from thermodynamic equilibrium, and the stability is determined by checking the invariance of concentrations. But for a closed system, the extent of reaction, ξ, can be expressed in terms of moles of each species, N_A, N_B, N_C and N_D and the volume of the system, V.

$$\frac{\Delta N_A}{-1} = \frac{\Delta N_B}{-1} = \frac{\Delta N_C}{1} = \frac{\Delta N_D}{1} = \mathrm{d}\xi. \tag{6.10}$$

Now, for example,

$$\frac{\Delta N_A}{V} = \Delta[A] = \frac{-\mathrm{d}\xi}{V}, \tag{6.11}$$

where $[A]$ is the concentration of A. We assume that V is constant, and define,

$$y \equiv \frac{\xi}{V}. \tag{6.12}$$

The extent of reaction is zero when time $t = 0$. Hence, we write,

$$\int \mathrm{d}[A] = -\int \mathrm{d}y = y - 0, \tag{6.13}$$

and

$$[A] - [A]_0 = -y, \tag{6.14}$$

where $[A]_0$ is the concentration at $t = 0$. We can write the following ODE for the time evolution of $[A]$:

$$-\frac{\mathrm{d}y}{\mathrm{d}t} = \frac{\mathrm{d}[A]}{\mathrm{d}t} = -k_{\mathrm{f}}[A][B] + k_{\mathrm{r}}[C][D]. \tag{6.15}$$

For ease of notation, we denote the initial values of concentration as,

$$[A]_0 = a,\ [B]_0 = b,\ [C]_0 = c,\ \text{and } [D]_0 = d.$$

Hence,

$$\frac{dy}{dt} = \gamma y^2 + \alpha y + \beta, \quad y(0) = 0. \tag{6.16}$$

Here

$$\alpha \equiv (c + d) - (a + b);$$

$$\beta \equiv k_f ad + k_r cd; \text{ and}$$

$$\gamma \equiv k_f + k_r.$$

At NESS,

$$\frac{dy}{dt} = 0 \Rightarrow \gamma y^2 + \alpha y + \beta = 0. \tag{6.17}$$

Two solutions are:

$$y_1 = \frac{\left(-\alpha - \sqrt{\alpha^2 - 4\beta\gamma}\right)}{2\gamma}; \tag{6.18}$$

and

$$y_2 = \frac{\left(-\alpha + \sqrt{\alpha^2 - 4\beta\gamma}\right)}{2\gamma}. \tag{6.19}$$

For a real solution to exist:

$$\lambda \equiv \alpha^2 - 4\beta\gamma \geq 0. \tag{6.20}$$

We can calculate the two solutions of the extent of reaction per unit volume (y_1, y_2) but the valid solution must provide nonnegative concentration values. Depending on the coefficients of Eq. (6.16), the stationary states may differ significantly.

Let us consider the system containing the reaction where the system is maintained out of equilibrium by controlling inflows and out flows. Then we can express the excess entropy production in terms of affinity and the velocity of reactions: affinity,

$$\mathrm{A} = RT \ln\left(\frac{R_f}{R_r}\right), \tag{6.21}$$

velocity,

$$\nu = (R_f - R_r), \tag{6.22}$$

and the excess entropy production,

$$\frac{\mathrm{d}\delta^2 S}{\mathrm{d}t} = 2\int \sum_k \delta F_k \delta J_k \mathrm{d}V. \tag{6.23}$$

Since we have a single reaction, $\delta F = \frac{\delta \mathrm{A}}{T}$ and $\delta J = \delta \nu$. Now

$$\mathrm{A} = RT \ln\left(\frac{k_\mathrm{f}[A][B]}{k_\mathrm{r}[C][D]}\right), \tag{6.24}$$

and

$$\nu = k_\mathrm{f}[A][B] - k_\mathrm{r}[C][D]. \tag{6.25}$$

Let us assume that the system is perturbed at a steady state by δB keeping other species at their stationary values. The stationary values are denoted by subscript s. Hence

$$[B] = [B]_\mathrm{s} + \delta B. \tag{6.26}$$

Then

$$\delta \mathrm{A} = RT \ln\left(\frac{k_\mathrm{f}[A]_\mathrm{s}([B]_\mathrm{s} + \delta B)}{k_\mathrm{f}[C]_\mathrm{s}[D]_\mathrm{s}}\right), \tag{6.27}$$

and

$$\delta \nu = k_\mathrm{f} \delta B [A]_\mathrm{s}. \tag{6.28}$$

Hence,

$$\delta F = \frac{\delta \mathrm{A}}{T} = R \ln\left(\frac{k_\mathrm{f}[A]_\mathrm{s}([B]_\mathrm{s} + \delta B)}{k_\mathrm{r}[C]_\mathrm{s}[D]_\mathrm{s}}\right), \tag{6.29}$$

and

$$\delta J = \delta \nu = k_\mathrm{f} \delta B [A]_\mathrm{s}. \tag{6.30}$$

We can write

$$\begin{aligned} \frac{\mathrm{d}\delta^2 S}{\mathrm{d}t} &= 2\int \sum_k \delta F_k \delta J_k \mathrm{d}V \\ &= 2\int R \ln\left(\frac{k_\mathrm{f}[A]_\mathrm{s}([B]_\mathrm{s} + \delta B)}{k_\mathrm{r}[C]_\mathrm{s}[D]_\mathrm{s}}\right) k_\mathrm{f} \delta B [A]_\mathrm{s} \mathrm{d}V. \end{aligned} \tag{6.31}$$

But at a critical point, reaction velocity is zero that gives

$$k_f[A]_s[B]_s = k_r[C]_s[D]_s. \tag{6.32}$$

Substituting back to Eq. (6.31), we obtain the excess entropy production as,

$$\frac{d\delta^2 S}{dt} = 2\int \sum_k \delta F_k \delta J_k dV = 2\int R \ln\left(1 + \frac{\delta B}{[B]_s}\right) k_f \delta B[A]_s dV. \tag{6.33}$$

For $\delta B \ll [B]_s$,

$$\ln\left(1 + \frac{\delta B}{[B]_s}\right) \cong \frac{\delta B}{[B]_s}.$$

Hence,

$$\frac{d\delta^2 S}{dt} = 2k_f \int \frac{[A]_s}{[B]_s} \delta B^2 dV > 0. \tag{6.34}$$

Therefore, the critical point under the perturbation of $[B]$ is stable. The effect of this perturbation can be explored using Eq. (6.16). If

$$\lambda \equiv \alpha^2 - 4\beta\gamma \geq 0$$

there are real critical points, we can start from a critical point, and considering a stationary point of y, μ, we can express a point very near to μ as,

$$y(t) = \mu + \delta y, \tag{6.35}$$

then substituting back to Eq. (6.16), we obtain,

$$\frac{d(\mu + \delta y)}{dt} = \gamma(\mu + \delta y)^2 + \alpha(\mu + \delta y) + \beta, \tag{6.36}$$

and as μ is a constant, $\delta y(0)$ is the perturbation at $t = 0$, i.e., the time starts at the moment the system is perturbed. By simplifying Eq. (6.36), we obtain,

$$\frac{d(\delta y)}{dx} = \gamma\delta y^2 + (2\gamma\mu + \alpha)\delta y + \left(\gamma\mu^2 + \alpha\mu + \beta\right). \tag{6.37}$$

We can solve Eq. (6.37) numerically to understand the behavior of δy. Note that since y is the extent of reaction per unit volume, it could be negative. Since the excess entropy production is positive for any perturbation of B, we expect the critical point to be stable, validating thermodynamic analysis via numerical solutions.

The computations related to the above analysis are given in the Mathematica® program (Appendix). We use $\{a = 1; b = 0.8; c = 0; d = 0; k_f = 0.5; k_r = 0.3\}$ for the sample computation, and for these values $\lambda = 1.96$, and there are two critical points at $y = 0.25$, and $y = 2.0$, but given the initial concentration values, the latter value may not occur frequently, hence we choose $y = 0.25$, and when we plot y's time evolution, it reaches $y = 0.25$ asymptotically over time. This reaction is stable for many sets of coefficients as the reader can test using the Mathematica® code provided to plot the graphs.

The same reaction is modeled for an open system in such a way that the concentration of C is kept constant at c. For ease of notation, we denote $[A] = x$, $[C] = c$, a constant, and the initial values of concentration of A, B, C, and D are denoted as a, b, c, and d, respectively. We numerically solve

$$\frac{\mathrm{d}x[t]}{\mathrm{d}t} = -k_f x[t](x[t] - a + b) + k_r c(-x[t] + a), \tag{6.38}$$

where for notational convenience, $x \equiv [A]$. Note that x is different to y in Eq. (6.16) above conceptually. We can also show that writing ODEs for $[B]$ and $[D]$,

$$[B] = x[t] - a + b, \tag{6.39}$$

and

$$[D] = -x[t] + a, \tag{6.40}$$

and $[C]$ is kept constant at c. We can recast Eq. (6.38) to the following:

$$\frac{\mathrm{d}x}{\mathrm{d}t} = \gamma x^2 + \alpha x + \beta, \tag{6.41}$$

where

$$\alpha = k_f(a - b) - k_r c, \tag{6.42}$$

$$\beta = k_r c a, \tag{6.43}$$

and

$$\gamma = -k_f. \tag{6.44}$$

The analytical solutions to Eq. (6.41) are given by,

$$x_1 = \left(-\alpha - \left[\sqrt{\alpha^2 - 4\beta\gamma}\right]\right)/(2\gamma), \tag{6.45}$$

and

$$x_2 = \left(-\alpha + \sqrt{\alpha^2 - 4\beta\gamma}\right)/(2\gamma). \tag{6.46}$$

Note that for real solutions,

$$\lambda \equiv \alpha^2 - 4\beta\gamma \geq 0.$$

Following symmetry, we can see that

$$\frac{d\delta^2 S}{dt} = 2k_f \int \frac{a}{b}\delta x^2 dV > 0 \tag{6.47}$$

when we perturb $[A]$ by $\delta A \equiv \delta x$. Hence the system is stable under perturbation near a NESS.

If $\lambda \equiv \alpha^2 - 4\beta\gamma \geq 0$, there are real critical points, we can start from a critical point, and considering a stationary point of x, μ, we can express a point very near to μ as, $x(t) = \mu + \delta x$, then substituting back to Eq. (6.41), we obtain,

$$\frac{d(\mu + \delta x)}{dt} = \gamma(\mu + \delta x)^2 + \alpha(\mu + \delta x) + \beta, \tag{6.48}$$

and as μ is a constant, $\delta x(0)$ is the perturbation at $t = 0$, i.e., the time starts at the moment the system is perturbed. By simplifying Eq. (6.48), we obtain,

$$\frac{d(\delta x)}{dx} = \gamma\delta x^2 + (2\gamma\mu + \alpha)\delta x + \left(\gamma\mu^2 + \alpha\mu + \beta\right). \tag{6.49}$$

We can solve Eq. (6.49) numerically to understand the behavior of δx.

The Mathematica® code for the open system analysis is also given in Appendix. By running the program, one can solve Eq. (6.49) for the concentrations of A, B, and D to see that they all reach steady states for the coefficients, for example, $\{a = 1; b = 0.8; c = 0.2; d = 0; k_f = 0.5; k_r = 0.3\}$. Stationary value for $[A] = 0.388$. Then we solve the perturbation ODE numerically for two different initial perturbations to see that the perturbations diminish over time reaching zero asymptotically. This confirms the entropy analysis by solving the perturbation ODEs numerically.

6.3 Example 2

Consider the autocatalytic reaction in an open system presented by Kondepudi and Prigogine (2014):

$$2X + Y \rightleftharpoons 3X \tag{6.50}$$

with the forward rate coefficient k_{f} and the reverse rate coefficient k_{r}. The forward reaction rate,

$$R_{\mathrm{f}} = k_{\mathrm{f}}[X]^2[Y], \tag{6.51}$$

the reverse reaction rate,

$$R_{\mathrm{r}} = k_{\mathrm{r}}[X]^3, \tag{6.52}$$

affinity,

$$\mathrm{A} = RT \ln\left(\frac{R_{\mathrm{f}}}{R_{\mathrm{r}}}\right), \tag{6.53}$$

and velocity,

$$\nu = R_{\mathrm{f}} - R_{\mathrm{r}}. \tag{6.54}$$

Then,

$$\nu = k_{\mathrm{f}}[X]^2[Y] - k_{\mathrm{r}}[X]^3. \tag{6.55}$$

Note that at a NESS,

$$\frac{\mathrm{d}[X]}{\mathrm{d}t} = -k_{\mathrm{f}}[X]^2[Y] + k_{\mathrm{r}}[X]^3 = 0 \tag{6.56}$$

making

$$k_{\mathrm{f}}[X]_{\mathrm{s}}^2[Y]_{\mathrm{s}} = k_{\mathrm{r}}[X]_{\mathrm{s}}^3, \tag{6.57}$$

where subscript s denotes the stationary values. Equation (6.57) gives,

$$\frac{[Y]_{\mathrm{s}}}{[X]_{\mathrm{s}}} = \frac{k_{\mathrm{r}}}{k_{\mathrm{f}}}. \tag{6.58}$$

From the above Eqs. (6.53), (6.55), and (6.58), we obtain,

$$\mathrm{A} = RT \ln\left(\frac{k_{\mathrm{f}}[Y]}{k_{\mathrm{r}}[X]}\right). \tag{6.59}$$

Suppose we perturb $[X]$ by an infinitesimal amount of δX at a NESS keeping $[Y]_{\mathrm{s}}$ at a constant value, then we write the perturbation in ν,

$$\delta\nu = k_{\mathrm{f}}([X]_{\mathrm{s}} + \delta X)^2([Y]_{\mathrm{s}}) - k_{\mathrm{r}}([X]_{\mathrm{s}} + \delta X)^3. \tag{6.60}$$

By using Eq. (6.58), we see that

$$\mathrm{A} = RT\ln\left(\frac{k_{\mathrm{f}}[Y]_{\mathrm{s}}}{k_{\mathrm{r}}[X]_{\mathrm{s}}}\right) = RT\ln(1) = 0$$

at a NESS. Now the excess entropy production can be written as,

$$\frac{\mathrm{d}\delta^2 S}{\mathrm{d}t} = 2\int\sum_k \delta F_k \delta J_k \mathrm{d}V = 2\int R\ln\left(\frac{k_{\mathrm{f}}[Y]_{\mathrm{s}}}{k_{\mathrm{r}}([X]_{\mathrm{s}}+\delta X)}\right)$$
$$\left(k_{\mathrm{f}}([X]_{\mathrm{s}}+\delta X)^2([Y]_{\mathrm{s}}) - k_{\mathrm{r}}([X]_{\mathrm{s}}+\delta X)^3\right)\mathrm{d}V. \tag{6.61}$$

In this case, the excess entropy production either positive or negative depending on k_{f} and k_{r}. Hence, we define

$$\theta e \equiv \ln\left(\frac{k_{\mathrm{f}}[Y]_{\mathrm{s}}}{k_{\mathrm{r}}([X]_{\mathrm{s}}+\delta X)}\right)\left(k_{\mathrm{f}}([X]_{\mathrm{s}}+\delta X)^2([Y]_{\mathrm{s}}) - k_{\mathrm{r}}([X]_{\mathrm{s}}+\delta X)^3\right) \tag{6.62}$$

to evaluate the excess entropy production numerically. If θe is negative or positive, then the excess entropy production is negative or positive, which indicates the instability or stability of the reaction, respectively. Equation (6.62) can be written as,

$$\theta e = \ln\left(\frac{1}{\left(1+\frac{\delta X}{a}\right)}\right)\left(k_{\mathrm{f}}(a+\delta X)^2 b - k_{\mathrm{r}}(a+\delta X)^3\right), \tag{6.63}$$

where we define $a \equiv [X]_{\mathrm{s}}$, and $b \equiv [Y]_{\mathrm{s}}$. For notational convenience, we define $x \equiv [X]$ and $y \equiv [Y]$ hereafter. Then we write the following ODEs, to describe the time evolution of x and y:

$$\frac{\mathrm{d}x}{\mathrm{d}t} = -k_{\mathrm{f}}x^2 y + k_{\mathrm{r}}x^3, \tag{6.64}$$

and

$$\frac{\mathrm{d}y}{\mathrm{d}t} = -k_{\mathrm{f}}x^2 y. \tag{6.65}$$

By making these derivatives zero, we see that the stationary point is a trivial solution, i.e., $x = 0$. Since this system is open, we can keep, for example, $y = c$, a constant; then, at NESS, $x = \frac{k_{\mathrm{f}}c}{k_{\mathrm{r}}}$. We can solve Eqs. (6.64) and (6.65) numerically as a system of ODEs and explore the behavior of the system.

Computations related to Example 2 are also given in the Mathematica® in Appendix. Readers can run the program for θe, and plot it to see that θe is very much dependent on the coefficients and in large regions of the coefficients, there are no positive values indicating that the system is unstable in these regions. It is also seen that in the numerical solutions of the system of ODEs that NESSs depend on the coefficients and the initial conditions, if there are any. In some regions of k_f and k_r, singularities exist. If we keep y constant at c, and when the stationary value for x is less than the initial value of x, x_0, we can derive a constraint for c $\left(c < \frac{k_f x_0}{k_r}\right)$ using Eq. (6.63), if there is a stationary value other than zero. Numerical solutions do not show stationary points. Hence, we can conclude that NESSs do not exist for the reaction except when x is zero, which is a trivial solution. By plotting time curves of x with several values of c, we see that at smaller values of c, solutions exhibit singularities.

6.4 Example 3

Consider the reaction scheme discussed in Kondepudi and Prigogine (2014), and readers can study the reaction scheme in more detail using their text. The purpose of our discussion here is to show the complexities inherent in such reaction schemes even without the presence of noise in the system. The reaction scheme is given below.

$$A \rightarrow^{k_1} X, \tag{6.66}$$

$$B + X \rightarrow^{k_2} Y + D, \tag{6.67}$$

$$2X + Y \rightarrow^{k_3} 3X, \tag{6.68}$$

and

$$X \rightarrow^{k_4} E. \tag{6.69}$$

All the reactions here are forward reactions with the rate coefficients indicated above the arrows. Given that the third reaction is in the scheme, which does not have stationary values based on the thermodynamic analysis shown previously, we could expect this reaction scheme can show very complex behaviors. Hence, we proceed to simplify the situation by assuming that D and E are removed from the system in such a way to maintain the concentrations of A, B, D, and E at constant values. Then this system reduces to a system of two variable concentrations, X and Y. Then we can write the following ODEs by using the notations $a \equiv [A]$, $b \equiv [B]$, which are constants, and $x \equiv [X]$, $y \equiv [Y]$.

$$\frac{dx}{dt} = k_1 a - k_2 b x + k_3 x^2 y - k_4 x = z_1, \tag{6.70}$$

and

$$\frac{dy}{dt} = k_2bx - k_3x^2y = z_2. \tag{6.71}$$

The stationary solutions can be obtained by making $z_1 = 0$, and $z_2 = 0$, and they are

$$x_s = \frac{k_1}{k_2}a, \tag{6.72}$$

and

$$y_s = \frac{k_2k_4}{k_1k_3}\frac{b}{a}. \tag{6.73}$$

These stationary values need to be checked for the system stability around them, and for that purpose we use local stability theory. The Jacobian matrix is defined at the stationary values as,

$$\Lambda = \begin{bmatrix} \frac{\partial z_1}{\partial x} & \frac{\partial z_1}{\partial y} \\ \frac{\partial z_2}{\partial x} & \frac{\partial z_2}{\partial y} \end{bmatrix} = \begin{bmatrix} -k_2b + 2k_3x_sy_s - k_4 & k_3x_s^2 \\ k_2b - 2k_3x_sy_s & -k_3x_s^2 \end{bmatrix}. \tag{6.74}$$

The Jacobian can be simplified using Eqs. (6.72) and (6.73),

$$\Lambda = \begin{pmatrix} k_2b - k_4 & k_3x_s^2 \\ -k_2b & -k_3x_s^2 \end{pmatrix}. \tag{6.75}$$

The stability of stationary states is investigated through the Eigen values of the characteristic equation:

$$\mathrm{Det}[\Lambda - \lambda I] = 0, \tag{6.76}$$

where λs are Eigen values and I is the identity matrix. Through this analysis, one can determine the conditions under which the stationary values become unstable, i.e., the real parts of complex conjugate pairs of the solutions of Eq. (6.76) are positive. The transition from stability to instability occurs when the real parts change from negative values to the positive values. This leads to the following condition for instability:

$$b > \frac{k_4}{k_2} + \frac{k_4}{k_2}\frac{k_1^2}{k_4^2}a^2. \tag{6.77}$$

The Mathematica® program given in Appendix shows how the system behaves near a transition point from stability to instability. The numerical solutions are shown within the stable region and then unstable region where the oscillatory behaviors exhibit themselves because of moving away from the stability. The reader is encouraged to use the program to test the behaviors by changing the coefficients. More mathematically inclined can check the conditions under which the system becomes a Hopf bifurcation.

Appendix: Mathematica® Program

(*Example 1: Chemical reaction $A+B \rightleftharpoons C+D$ with forward reaction rate coefficient k_f and reverse reaction rate k_r. First, we compute deterministic chemical reactions related to a closed system that reaches thermodynamic equilibrium. For notational simplicity, the initial values of concentration of *A*, *B*, *C*, and *D* are denoted as *a*, *b*, *c*, and *d*, respectively. *y* is the extent of reaction per unit volume. See the derivations in Chap. 6, and the outputs are not shown.*)

(*We solve the differential equation y′[t] = gamma y^2 + alpha y + beta for stationary states*)

Solve [gamma y^2 + alpha y + beta == 0, y]

$$\left\{\left\{y \to \frac{-\text{alpha} - \sqrt{\text{alpha}^2 - 4\,\text{beta gamma}}}{2\,\text{gamma}}\right\}, \left\{y \to \frac{-\text{alpha} + \sqrt{\text{alpha}^2 - 4\,\text{beta gamma}}}{2\,\text{gamma}}\right\}\right\}$$

$$y1 = \frac{-\text{alpha} - \sqrt{\text{lambda}}}{2\,\text{gamma}}$$

$$y2 = \frac{-\text{alpha} + \sqrt{\text{lambda}}}{2\,\text{gamma}}$$

$$\text{lambda} = \text{alpha}^2 - 4\,\text{beta gamma};$$

(*lambda needs to be nonnegative for *a* solution to exist*)

(*First we define alpha, beta, gamma, and lambda according to the equations given in the text*)

alpha = (c + d) − (a + b); beta = kf a b + kr c d; gamma = kf + kr;

(*Numerical evaluations*)

(*First check whether lambda is nonnegative*)

a = 1.0; b = 0.8; c = 0.0; d = 0.0; kf = 0.5; kr = 0.3;lambda

(*We solve the differential equation y′[t] = gamma y^2 + alpha y + beta for stationary states*)

(*Numerically solve the ODE for the given values*)

sol = NDSolve[{y′[t] == gamma y[t]^2 + alpha y[t] + beta, y[0] == 0}, y, {t, 0, 20}]

Plot [Evaluate[y[t] /. sol], {t, 0, 20}, PlotRange → All,

```
Axes → True, Frame → True, AxesLabel → Automatic,
PlotLabel → "Solution for a = 1.0; b = 0.8; c = 0.0; d = 0.0; kf = 0.5; kr = 0.3"]
(*Let us perturb y by deltay, a small value, and solve the ODE for the
perturbation*)
a = 1.0; b = 0.8; c = 0.0; d = 0.0; kf = 0.5; kr = 0.3;
alpha = (c + d) − (a + b); beta = kf a b + kr c d; gamma = kf + kr;
lambda = alpha^2 − 4 beta gamma;
mu = 0.25;
soldeltay = NDSolve[{deltay′[t] ==
    gamma deltay[t]^2 + (2 gamma mu + alpha) deltay[t] + (gamma mu^2 +
    alpha mu + beta), deltay[0] == 0.005}, deltay, {t, 0, 5}]
Plot[Evaluate[deltay[t] /. soldeltay], {t, 0, 5}, PlotRange → All,
Axes → True, Frame → True, AxesLabel → Automatic,
PlotLabel → "Pertubation a = 1.0; b = 0.8; c = 0.0; d = 0.0; kf = 0.5; kr = 0.3"]
(*Increase the initial condition of the perturbation*)
soldeltay =
NDSolve[{deltay′[t] ==
gamma deltay[t]^2 + (2 gamma mu + alpha) deltay[t] + (gamma mu^2 +
alpha mu + beta), deltay[0] == 0.05}, deltay, {t, 0, 5}]
Plot[Evaluate[deltay[t] /. soldeltay], {t, 0, 5}, PlotRange → All,
Axes → True, Frame → True, AxesLabel → Automatic,
PlotLabel → "Pertubation a = 1.0; b = 0.8; c = 0.0; d = 0.0; kf = 0.5; kr = 0.3"]
(*The reaction in Example 1 modeled for an open system in such a way the
concentration of C is kept constant at c. For ease of notation, we denote [A] = x, [C]
= c, a constant, and the initial values of concentration of A, B, C, and D are denoted
as a, b, c, and d, respectively*)
(*Solve the equation x′[t] = gamma x^2 + alpha x + beta, analytically for
stationary values for x*)
Solve[gamma x^2 + alpha x + beta == 0, x]
(*We define alpha, beta, gamma, and lambda according to the equations given in
the text*)
alpha = kf (a − b) − kr c; beta = kr c a; gamma = −kf; lambda =
alpha^2 − 4 beta gamma;
(*Numerical evaluations*)
a = 1.0; b = 0.8; c = 0.2; d = 0.0; kf = 0.5; kr = 0.3;
(*here c is kept as a constant*)
lambda
x1 = (− alpha − Sqrt[alpha^2 − 4 beta gamma])/(2 gamma)
x2 = (− alpha + Sqrt[alpha^2 − 4 beta gamma])/(2 gamma)
(*x1 is the only valid stationary value*)
(*Solve the ODE numerically*)
sol = NDSolve[{x′[t] == gamma x[t]^2 + alpha x[t] + beta, x[0] == 1.0},
x, {t, 0, 20}]
Plot [Evaluate[{x[t], (x[t] − a + b), (− x[t] + a)} /. sol], {t, 0, 20},
PlotRange → All, Axes → True, Frame → True,
```

```
AxesLabel → Automatic,
PlotLegends → {"[A]", "[B]", "[D]"},
PlotLabel →
"Concentration of A, B, D for a = 1.0; b = 0.8; c = 0.2; d = 0.0; kf = 0.5; kr =
0.3"]
(*The reaction in Example 1 modeled for an open system in such a way the
concentration of C is kept constant at c. For ease of notation, we denote [A] = x, [C]
= c, a constant, and the initial values of concentration of A, B, C, and D are denoted
as a, b, c, and d, respectively*)
(*Solve the equation x′[t] = gamma x^2 + alpha x + beta, analytically for
stationary values for x*)
Solve[gamma x^2 + alpha x + beta == 0, x]
(*We define alpha, beta, gamma, and lambda according to the equations given in
the text*)
alpha = kf (a − b) − kr c; beta = kr c a; gamma = −kf; lambda =
alpha^2 − 4 beta gamma;
(*Numerical evaluations*)
a = 1.0; b = 0.8; c = 0.2; d = 0.0; kf = 0.5; kr = 0.3;
(*here c is kept constant*)
lambda
x1 = (− alpha − Sqrt[alpha^2 − 4 beta gamma])/(2 gamma)
x2 = (− alpha + Sqrt[alpha^2 − 4 beta gamma])/(2 gamma)
(*x1 is the only valid stationary value*)
(*Solve the ODE numerically*)
sol = NDSolve[{x′[t] == gamma x[t]^2 + alpha x[t] + beta, x[0] == 1.0},
x, {t, 0, 20}]
Plot[Evaluate[{x[t], (x[t] − a + b), (− x[t] + a)} /. sol], {t, 0, 20},
PlotRange → All, Axes → True, Frame → True, AxesLabel → Automatic,
PlotLegends → {"[A]", "[B]", "[D]"},
PlotLabel →
"Concentration of A, B, D for a = 1.0; b = 0.8; c = 0.2; d = 0.0; kf = 0.5; kr =
0.3"]
(*Perturbation analysis: Perturb [A] by deltax, and solve the ODE taking t = 0
when perturbation starts*)
mu = 0.388; (*stationary value of [A]*)
soldeltax =
NDSolve[{deltax′[t] ==
gamma deltax[t]^2 + (2 gamma mu + alpha) deltax[t] + (gamma mu^2 +
alpha mu + beta), deltax[0] == 0.005}, deltax, {t, 0, 10}]
Plot[Evaluate[{deltax[t], (deltax[t]), (− deltax[t])} /. soldeltax], {t, 0,
10}, PlotRange → All, Axes → True, Frame → True,
AxesLabel → Automatic,
PlotLegends → {"delta[A]", "delta[B]", "delta[D]"},
PlotLabel →
```

```
"Perturbation of A, B, D for a = 1.0; b = 0.8; c = 0.2; d = 0.0; kf = 0.5; kr =
0.3"]
soldeltax =
NDSolve[{deltax'[t] ==
gamma deltax[t]^2 + (2 gamma mu + alpha) deltax[t] + (gamma mu^2 +
alpha mu + beta), deltax[0] == 0.01}, deltax, {t, 0, 10}]
Plot[Evaluate[{deltax[t], (deltax[t]), (− deltax[t])} /. soldeltax], {t, 0,
10}, PlotRange → All, Axes → True, Frame → True, AxesLabel → Automatic,
PlotLegends → {"delta[A]", "delta[B]", "delta[D]"},
PlotLabel →
"Perturbations of A, B, D for a = 1.0; b = 0.8; c = 0.2; d = 0.0; kf = 0.5; kr =
0.3"]
(*Example 2: The reaction 2x + y = 3x in an open system*)
(*Excess entropy production related function thetae*)
thetae[a_, b_, deltax_, kf_, kr_]: =
Log[1.0/(1 + (deltax/a))] (kf ((a + deltax)^2) b − kr (a + deltax)^3);
Plot3D[thetae[a, b, 0.001, 0.5, 0.4], {a, 0.01, 1.0}, {b, 0.01, 1.0},
PlotRange → All, ColorFunction → Hue]
(*Solving two ODEs numerically*)
kf = 0.5; kr = 0.4;
soln = NDSolve[{x'[t] == − kf (x[t]^2) y[t] + kr (x[t]^3),
y'[t] == − kf (x[t]^2) y[t], x[0] == 1.0, y[0] == 1.0}, {x, y}, {t, 0, 2.95}]
ParametricPlot[Evaluate[{x[t], y[t]} /. soln], {t, 0, 3},
PlotRange → Automatic, Axes → True, Frame → True,
PlotStyle → {{Black, Thick}, {Blue, Thick}}, AxesLabel → {"time"},
PlotLabel → "X and Y kf = 0.5, kr = 0.4, x0 = 1.0, y0 = 1.0"]
kf = 0.6; kr = 0.4;
soln = NDSolve[{x'[t] == − kf (x[t]^2) y[t] + kr (x[t]^3),
y'[t] == − kf (x[t]^2) y[t], x[0] == 1.0, y[0] == 1.0}, {x, y}, {t, 0, 3.5}]
ParametricPlot[Evaluate[{x[t], y[t]} /. soln], {t, 0, 3},
PlotRange → Automatic, Axes → True, Frame → True,
PlotStyle → {{Black, Thick}, {Blue, Thick}}, AxesLabel → {"time"},
PlotLabel → "X and Y kf = 0.5, kr = 0.4, x0 = 1.0, y0 = 1.0"]
Plot[Evaluate[{x[t], y[t]} /. soln], {t, 0, 2.95}, PlotRange → All,
Axes → True, Frame → True, PlotStyle → {{Black, Thick}, {Blue, Thick}},
AxesLabel → {"time"}, PlotLabel → "X and Y
kf = 0.5, kr = 0.4, x0 = 1.0, y0 = 1.0"]
kf = 0.5; kr = 0.4; c = 0.3;
soln = NDSolve[{x'[t] == − kf (x[t]^2) c + kr (x[t]^3), x[0] == 0.4},
x, {t, 0, 20}]
Plot[Evaluate[x[t] /. soln], {t, 0, 20}, PlotRange → All, Axes → True,
Frame → True, PlotStyle → {{Black, Thick}}, AxesLabel → {"time"},
PlotLabel → "X kf = 0.5, kr = 0.4, x0 = 0.4, c = 0.3"]
kf = 0.5; kr = 0.4; c = 0.4;
soln = NDSolve[{x'[t] == − kf (x[t]^2) c + kr (x[t]^3), x[0] == 0.4},
```

```
x, {t, 0, 20}]
Plot[Evaluate[x[t] /. soln], {t, 0, 20}, PlotRange → All,
Axes → True,
Frame → True, PlotStyle → {{Black, Thick}},
AxesLabel → {"time"},
PlotLabel → "X kf = 0.5, kr = 0.4, x0 = 0.4, c = 0.4"]
kf = 0.5; kr = 0.4; c = 1.0;
soln = NDSolve[{x′[t] == − kf (x[t]^2) c + kr (x[t]^3), x[0] == 0.4},
x, {t, 0, 20}]
Plot[Evaluate[x[t] /. soln], {t, 0, 20}, PlotRange → All, Axes → True,
Frame → True, PlotStyle → {{Black, Thick}}, AxesLabel → {"time"},
PlotLabel → "X kf = 0.5, kr = 0.4, x0 = 0.4, c = 1.0"]
kf = 0.5; kr = 0.4; c = 0.8;
soln = NDSolve[{x′[t] == − kf (x[t]^2) c + kr (x[t]^3), x[0] == 0.4},
x, {t, 0, 20}]
Plot[Evaluate[x[t] /. soln], {t, 0, 20}, PlotRange → All, Axes → True,
Frame → True, PlotStyle → {{Black, Thick}}, AxesLabel → {"time"},
PlotLabel → "X kf = 0.5, kr = 0.4, x0 = 0.4, c = 0.8"]
kf = 0.5; kr = 0.4; c = 0.2;
soln = NDSolve[{x′[t] == − kf (x[t]^2) c + kr (x[t]^3), x[0] == 0.4},
x, {t, 0, 14}]
Plot[Evaluate[x[t] /. soln], {t, 0, 14}, PlotRange → All,
Axes → True, Frame → True, PlotStyle → {{Black, Thick}},
AxesLabel → {"time"},
PlotLabel → "X kf = 0.5, kr = 0.4, x0 = 0.4, c = 0.2, 0.8 and 1.0"]
soln = NDSolve[{x′[t] == − kf (x[t]^2) c + kr (x[t]^3), x[0] == 0.4},
x, {t, 0, 14}]
Plot[Evaluate[x[t] /. soln], {t, 0, 14}, PlotRange → All, Axes → True,
Frame → True, PlotStyle → {{Black, Thick}}, AxesLabel → {"time"},
PlotLabel → "X kf = 0.5, kr = 0.4, x0 = 0.4, c = 0.2, 0.8 and 1.0"]
(*Example 3: The BrusselatorKondepudi and Prigogine (2014)*)
k1 = 1.0; k2 = 1.0; k3 = 1.0; k4 = 1.0; A = 1.0; B = 3.0; R1 := k1 * A;
R2 : = k2 * B * X[t]; R3: = k3 * (X[t]^2) * Y[t];
R4: = k4 * X[t];
lambda := (k4/k2) + ((k3 k1^2)/(k2 k4^2)) A^2;
lambda
2.
(*B > 2, hence the stationary states are unstable*)
(*Stationary states are computed*)
Xs = (k1/k2) A
Ys = ((k4 k2)/(k3 k1)) (B/A)
soln = NDSolve[{X′[t] == R1 − R2 + R3 − R4, Y′[t] == R2 − R3,
X[0] == 1.0, Y[0] == 1.0}, {X, Y}, {t, 0, 20}, MaxSteps → 500]
Plot[Evaluate[{X[t]} /. soln], {t, 0, 20}, PlotRange → {0, 4},
PlotStyle → {Black, Thick}, AxesLabel → {"time"},
```

```
PlotLabel → "X for k1 = 1.0; k2 = 1.0; k3 = 1.0; k4 = 1.0; A = 1.0; B = 3.0",
AxesStyle → Directive[Black, 14]]
Plot[Evaluate[{X[t], Y[t]} /. soln], {t, 0, 20}, PlotRange → {0, 5},
PlotStyle → {{Black, Thick}, {Gray, Thick}}, AxesLabel → {"time"},
PlotLabel → "X and Y for k1 = 1.0; k2 = 1.0; k3 = 1.0; k4 = 1.0; A = 1.0; B
= 3.0",
AxesStyle → Directive[Black, 14]]
(*Let us change B to 1.5*)
B = 1.5;
lambda
(*B < 2, hence the system is stable; stationary points are given below*)
Xs = (k1/k2) A
Ys = ((k4 k2)/(k3 k1)) (B/A)
soln = NDSolve[{X'[t] == R1 − R2 + R3 − R4, Y'[t] == R2 − R3, X[0] ==
1.0,
Y[0] == 1.0}, {X, Y}, {t, 0, 20}, MaxSteps → 500]
Plot[Evaluate[{X[t]} /. soln], {t, 0, 20}, PlotRange → {0, 2},
PlotStyle → {Black, Thick}, AxesLabel → {"time"},
PlotLabel → "X for k1 = 1.0; k2 = 1.0; k3 = 1.0; k4 = 1.0; A = 1.0; B = 1.5",
AxesStyle → Directive[Black, 14]]
Plot[Evaluate[{X[t], Y[t]} /. soln], {t, 0, 20}, PlotRange → {0, 2},
PlotStyle → {{Black, Thick}, {Gray, Thick}}, AxesLabel → {"time"},
PlotLabel → "X and Y for k1 = 1.0; k2 = 1.0; k3 = 1.0; k4 = 1.0; A = 1.0; B
= 1.5",
AxesStyle → Directive[Black, 14]]
(*Change B closer to 2 but less than 2*)
B = 1.9999;
k1 = 1.0; k2 = 1.0; k3 = 1.0; k4 = 1.0; A = 1.0; R1 := k1 * A; R2 := k2 * B *
X[t]; R3 := k3 * (X[t]^2) * Y[t];
R4 : = k4 * X[t];
soln = NDSolve[{X'[t] == R1 − R2 + R3 − R4, Y'[t] == R2 − R3, X[0] ==
1.0,
Y[0] == 1.0}, {X, Y}, {t, 0, 20}, MaxSteps → 500]
Plot[Evaluate[{X[t]} /. soln], {t, 0, 20}, PlotRange → {0, 2},
PlotStyle → {Black, Thick}, AxesLabel → {"time"},
PlotLabel → "X for k1 = 1.0; k2 = 1.0; k3 = 1.0; k4 = 1.0; A = 1.0; B =
1.9999",
AxesStyle → Directive[Black, 14]]
Plot[Evaluate[{X[t], Y[t]} /. soln], {t, 0, 20}, PlotRange → {0, 3},
PlotStyle → {{Black, Thick}, {Gray, Thick}}, AxesLabel → {"time"},
PlotLabel → "X and Y for k1 = 1.0; k2 = 1.0; k3 = 1.0; k4 = 1.0; A = 1.0; B
= 1.5",
AxesStyle → Directive[Black, 14]]
(*B = 2*)
B = 2.0;
```

```
Xs = (k1/k2) A
Ys = ((k4 k2)/(k3 k1)) (B/A)
soln = NDSolve[{X'[t] == R1 − R2 + R3 − R4, Y'[t] == R2 − R3, X[0] ==
1.0,
Y[0] == 1.0}, {X, Y}, {t, 0, 20}, MaxSteps → 500]
Plot[Evaluate[{X[t]} /. soln], {t, 0, 20}, PlotRange → {0, 2},
PlotStyle → {Black, Thick}, AxesLabel → {"time"},
PlotLabel → "X for k1 = 1.0; k2 = 1.0; k3 = 1.0; k4 = 1.0; A = 1.0; B = 2",
AxesStyle → Directive[Black, 14]]
Plot[Evaluate[{X[t], Y[t]} /. soln], {t, 0, 20}, PlotRange → {0, 3},
PlotStyle → {{Black, Thick}, {Gray, Thick}}, AxesLabel → {"time"},
PlotLabel → "X and Y for k1 = 1.0; k2 = 1.0; k3 = 1.0; k4 = 1.0; A = 1.0; B
= 2",
AxesStyle → Directive[Black, 14]]
(*Plot X and Y, time as a parameter*)
k1 = 1.0; k2 = 1.0; k3 = 1.0; k4 = 1.0; A = 1.0; R1 := k1 * A; R2 :=
k2 * B * X[t]; R3 := k3 * (X[t]^2) * Y[t];
R4 : = k4 * X[t]; B = 2.5;
soln = NDSolve[{X'[t] == R1 − R2 + R3 − R4, Y'[t] == R2 − R3, X[0] ==
1.0,
Y[0] == 1.0}, {X, Y}, {t, 0, 20}, MaxSteps → 500]
ParametricPlot[Evaluate[{X[t], Y[t]} /. soln], {t, 0, 20},
PlotRange → {0, 4}, PlotStyle → {{Black, Thick}, {Gray, Thick}},
AxesLabel → {"X", "Y"},
PlotLabel →
"Parametric Plot of X and Y for k1 = 1.0; k2 = 1.0; k3 = 1.0; k4 = 1.0; A = 1.0;
\
B = 2.5", AxesStyle → Directive[Black, 14]]
ParametricPlot[Evaluate[{X[t], Y[t]} /. soln], {t, 0, 300},
PlotRange → {0, 4}, PlotStyle → {{Black, Thick}, {Gray, Thick}},
AxesLabel → {"X", "Y"},
PlotLabel →
"Parametric Plot of X and Y for k1 = 1.0; k2 = 1.0; k3 = 1.0; k4 = 1.0; A = 1.0;
B = 2.5", AxesStyle → Directive[Black, 14]]
Plot[Evaluate[{X[t], Y[t]} /. soln], {t, 0, 300}, PlotRange → {0, 4},
PlotStyle → {{Black, Thick}, {Gray, Thick}}, AxesLabel → {"time"},
PlotLabel → "X and Y for k1 = 1.0; k2 = 1.0; k3 = 1.0; k4 = 1.0; A = 1.0; B
= 2.5",
AxesStyle → Directive[Black, 14]]
```

Reference

Kondepudi, D., & Prigogine, I. (2014). *Modern thermodynamics: From heat engines to dissipative structures* (2nd ed.). https://doi.org/10.1002/9781118698723

Chapter 7
Numerical Investigations of WSDEs in Chemical Transformations

7.1 Introduction

Even though the laboratories provide controlled environments for experiments without any significant noise except perhaps for instrumental errors, most of the chemical reactions that happen in natural settings as in life sciences occur in noisy environments. Noise would be externally induced or intrinsic, stemming from random collisions of smaller molecular numbers involved in a reaction. It is important to understand how noise influences the phenomena at hand. Therefore, it is imperative to model the situation appropriately. Here we assume that the noise that enters into a system is white, and WSDEs are the appropriate model to describe the effects of noise in a reaction. We also assume that noise is multiplicative, which is a reasonable assumption in many situations given that higher the concentration of a species more significant is the noise.

In this chapter, we mainly focused on closed systems, but one open system reaction is considered as the last example. The main emphasis of the Mathematica® codes is to provide the algorithms so that readers can use them to experiment with them in their own ways; whenever necessary, we show the outputs in the text.

7.2 A Simple Reaction: $X \rightleftharpoons Y$

7.2.1 Formulation of the Numerical Scheme

Consider the reaction

$$X \rightleftharpoons Y \tag{7.1}$$

D. Kulasiri, *Stochastic Differential Equations for Chemical Transformations in White Noise Probability Space*, https://doi.org/10.1007/978-981-97-9392-1_7

with forward reaction rate coefficient k_f and reverse rate coefficientk_r in a system closed for mass exchange with outside. Let us assume that the initial concentrations are $[X]_0 = x_0$ and $[Y]_0 = y_0$. To simplify the notation, we use $[X] = x$ and $[Y] = y$. Since the total mass of the system is conserved, we define,

$$\lambda_0 \equiv x_0 + y_0. \tag{7.2}$$

Hence, at any given time,

$$\lambda_0 = x + y, \text{ or}$$

$$y = \lambda_0 - x. \tag{7.3}$$

We also assume that k_f and k_r are irregular, i.e., noisy, and can be modeled by positive noise functions. Implicit in using the WSDEs is the use of white noise probability space as discussed in previous chapters. The following WSDE can be written for this reaction:

$$\begin{aligned} \frac{dx(t,\omega)}{dt} &= -k_f(t,\omega)\Diamond x(t,\omega) + k_r(t,\omega)\Diamond y(t,\omega) \\ &\quad + \alpha x(t,\omega)\Diamond w(t,\omega), \end{aligned} \tag{7.4}$$

where α is a constant coefficient acting as an amplitude for the white noise $w(t, \omega)$. Equation (7.5) is written using Eq. (7.4) and with simpler notation:

$$\frac{dx}{dt} = -k_f\Diamond x + k_r\Diamond(\lambda_0 - x) + \alpha x\Diamond w\cdot \tag{7.5}$$

In Eq. (7.5), we note that the fluctuations of x and y due to external and internal sources have a constraint based on mass conservation at each instant. We also express the rate coefficients as positive noise stochastic functions:

$$k_f(t,\omega) = k_{f0}\exp^{\Diamond}\left[a_f w(t,\omega)\right], \text{ and} \tag{7.6}$$

$$k_r(t,\omega) = k_{r0}\exp^{\Diamond}[a_r w(t,\omega)], \tag{7.7}$$

where a_f and a_r are amplitudes for $w(t, \omega)$.

We note in Chap. 4, upon discretization of t and the beginning of ith interval if we take $t = t_i$ as a constant, then we can approximate,

$$k_f(t_i,\omega) \approx k_{f0}\exp\left[a_f w(t_i,\omega)\right], \text{ and} \tag{7.8}$$

$$k_r(t_i,\omega) \approx k_{r0}\exp[a_r w(t_i,\omega)].$$

This approach would avoid the problem of decreasing time dependency of the Wick exponential (See Chap. 4) but give positive noise term. Here, we note that for a stochastic function F,

$$E\left[\exp^{\diamond} F\right] = \exp[E[F]]. \tag{7.9}$$

The expectation of $\frac{k_f(t,\omega)}{k_{f0}}$ would be,

$$E\left[\exp^{\diamond}\left[a_f w(t,\omega)\right]\right] = \exp\left[E\left[a_f w(t,\omega)\right]\right] = \exp[0] = 1\cdot \tag{7.10}$$

Assuming k_{f0} is a constant, then

$$E\left[k_f(t,\omega)\right] = k_{f0}\cdot \tag{7.11}$$

Similarly,

$$E[k_r(t,\omega)] = k_{r0}, \tag{7.12}$$

again, assuming that k_{r0} is a constant. This result justifies our models for stochastic rate coefficients with a positive noise. Here, a_f and a_r are factors that control the white noise as amplitudes, which can be used to adjust or estimate the noise content in the coefficients using experimental results, if available. Further, if we make $w(0,\omega) = 0$ at the beginning of a simulation, then the initial values of $k_f(0,\omega)$ and $k_f(0,\omega)$ are k_{f0} and k_{r0}, respectively. This means that the coefficients fluctuate around their respective expected values, where fluctuations are under the influence of a_f and a_r. Let us take the expectation of Eq. (7.5):

$$E\left[\frac{dx}{dt}\right] = -E\left[k_f \diamond x\right] + E[k_r \diamond (\lambda_0 - x)] + E[\alpha x \diamond w], \tag{7.13}$$

and if we write x in terms of Clarke-Ocone formula,

$$x = E[x] + \int_0^t \psi(t,\omega) dW(t,\omega), \tag{7.14}$$

where $\psi(t,\omega)$ is the Malliavin derivative of x, and W is the Wiener process. Let us define

$$E[x] \equiv f(t), \text{ and} \tag{7.15}$$

$$s(t,\omega) \equiv \int_0^t \psi(t,\omega) dW(t,\omega). \tag{7.16}$$

Note f is a deterministic function. This formulation is suitable for many physical and chemical systems, where we can measure a mean value from many experiments for a given time period under the same experimental conditions. This means that randomness is superimposed on the expected value, i.e., in the absence of noise we can expect a deterministic function as a solution. However, this approach is not suitable in situations where the expectation of a stochastic variable does not provide any meaningful interpretation within the context of experiments. Before expanding Eq. (7.5) using Eq. (7.14), let us write Eq. (7.13), using the relationship for two stochastic functions, F_1 and F_2,

$$E[F_1 \diamond F_2] = E[F_1]E[F_2], \tag{7.17}$$

$$\left(\frac{df}{dt} = -k_{f0}f + k_{r0}(\lambda_0 - f), f(0) = x_0\right). \tag{7.18}$$

By integrating using the separation of variable approach, we obtain,

$$f(t) = \left(\frac{k_{r0}\lambda_0}{k_{f0} + k_{r0}}\right) + \left(x_0 - \left(\frac{k_{r0}\lambda_0}{k_{f0} + k_{r0}}\right)\right)\exp\left[-\left(k_{f0} + k_{r0}\right)t\right] \tag{7.19}$$

Equation (7.19) is an exponentially decaying function with k_{f0} and k_{r0} playing major roles in the solution along with λ_0, total number of moles per unit volume in the system. We also may solve Eq. (7.18) numerically. By substituting Eq. (7.14) into Eq. (7.5), we obtain,

$$\frac{df}{dt} + \frac{ds}{dt} = -k_f \diamond (f + s) + k_r \diamond (\lambda_0 - f - s) + \alpha \diamond (f + s) \diamond w \tag{7.20}$$

and since α is deterministic, we write Eq. (7.20) as

$$\frac{df}{dt} + \frac{ds}{dt} = -k_f \diamond (f + s) + k_r \diamond (\lambda_0 - f - s) + \alpha (f + s) \diamond w. \tag{7.21}$$

In Eq. (7.21), we note that f is a deterministic function,
λ_0 is a constant, and all others are stochastic functions. Then we expand Eq. (7.21),

$$\frac{df}{dt} + \frac{ds}{dt} = -(k_f + k_r)f - (k_f + k_r)\diamond s + k_r\lambda_0 + \alpha f w + \alpha s \diamond w. \tag{7.22}$$

By using Eq. (7.18) to replace $\frac{df}{dt}$ in Eq. (7.22), we obtain,

$$\frac{ds}{dt} = (k_r - k_{r0})\lambda_0 - \left((k_f - k_{f0}) - (k_r - k_{r0})\right)f \\ - \left(k_f + k_r\right)\diamond s + \alpha f w + \alpha s \diamond w, \tag{7.23}$$

with the initial condition, $s(t = 0, \omega) = 0$.

We develop a finite difference scheme based on Eq. (7.23) by writing a difference equation for the time interval, $\Delta t_i = t_{i+1} - t_i$. When $t_i \leq t < t_{i+1}$,

$$\frac{(s_{i+1} - s_i)}{\Delta\, t_i} \cong (k_{ri} - k_{r0})\lambda_0 - \big((k_{fi} - k_{f0}) - (k_{ri} - k_{r0})\big)f_i \\ - \big(k_{fi} + k_{ri}\big)\Diamond s_i + \alpha f_i w_i + \alpha s_i \Diamond w_i. \tag{7.24}$$

We denote the values of stochastic and deterministic functions at t_i using the subscript i. By multiplying Eq. (7.24) by Δt_i throughout, we obtain,

$$(s_{i+1} - s_i) \cong (k_{ri} - k_{r0})\lambda_0 \Delta t_i - \big((k_{fi} - k_{f0}) - (k_{ri} - k_{r0})\big)f_i \Delta t_i \\ - \big(k_{fi} + k_{ri}\big)\Diamond s_i \Delta t_i + \alpha f_i w_i \Delta t_i + \alpha s_i \Diamond w_i \Delta t_i \tag{7.25}$$

In right hand side of Eq. (7.25) we note that all the functions are evaluated at the beginning of the time interval, and since Δt_i is a chosen value, we write Eq. (7.25) as,

$$s_{i+1} \cong s_i + (k_{ri} - k_{r0})\lambda_0 \Delta t_i - \big((k_{fi} - k_{f0}) - (k_{ri} - k_{r0})\big)f_i \Delta t_i \\ - \big(k_{fi} + k_{ri}\big)\Diamond s_i \Delta t_i + \alpha f_i \Delta W_i + \alpha s_i \Diamond \Delta W_i, \tag{7.26}$$

where W_i is the Wiener process at t_i. In the limit, $\Delta t_i \to 0$, we write,

$$s_{i+1} \cong s_i + (k_{ri} - k_{r0})\lambda_0 dt_i - \big((k_{fi} - k_{f0}) - (k_{ri} - k_{r0})\big)f_i dt_i \\ - \big(k_{fi} + k_{ri}\big)\Diamond s_i dt_i + \alpha f_i dW_i + \alpha s_i \Diamond dW_i. \tag{7.27}$$

If we divide the time interval [0, t] into equidistant infinitesimal time intervals of dt, then,

$$s_{i+1} \cong s_i + (k_{ri} - k_{r0})\lambda_0 dt - \big((k_{fi} - k_{f0}) - (k_{ri} - k_{r0})\big)f_i dt \\ - \big(k_{fi} + k_{ri}\big)\Diamond s_i dt + \alpha f_i dW_i + \alpha s_i \Diamond dW_i, \tag{7.28}$$

with the starting value of $s_0 = 0$. We make use of Eq. (7.28) in evaluating this scheme. Finally,

$$x_i = f_i + s_i. \tag{7.29}$$

The accuracy of this scheme can be improved by using smaller dt values. The Mathematica® code for this scheme is given in Appendix 7.1.

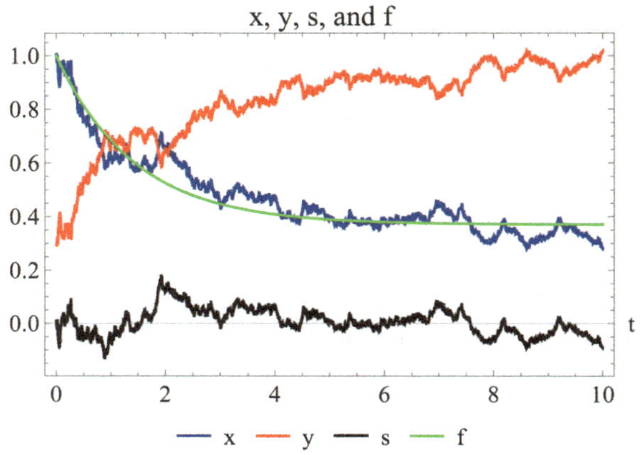

Fig. 7.1 A realization of x, y, f, and s

7.2.2 *Computations*

As seen in the Mathematica® code given in Appendix 7.1, first we compute the numerical scheme for a single realization using the values,

$$\{k_{f0} = 0.5;\ a_f = 3.;\ \alpha = 0.2;\ k_{r0} = 0.2;\ a_r = 3.;\\ \Delta t = 0.001;\ x_0 = 1.0;\ y_0 = 0.3\}.$$

The program has been tested using a range of coefficient values for its robustness. Figure 7.1 shows the outcomes for the set of coefficients and initial values given above. The plot of x follows that of f as expected. The stochastic function s fluctuates around zero providing the random component of x.

We compute 50 realizations of the variables to obtain the noise ratio of x, defined as the square root of variance of x divided by the mean of x at each time point. Then we repeat the same procedure by doubling α to 0.4 and keeping all the other coefficients the same. Figure 7.2 shows the two noise ratio curves: when α is doubled from 0.2, the noise ratio is increased by a factor greater than 2 at each time instant, which shows the effect of noise on the behavior of x.

7.3 Reaction $x + y \rightleftharpoons z$

7.3.1 *Formulation of the Numerical Scheme*

Consider the reaction

$$x + y \rightleftharpoons z$$

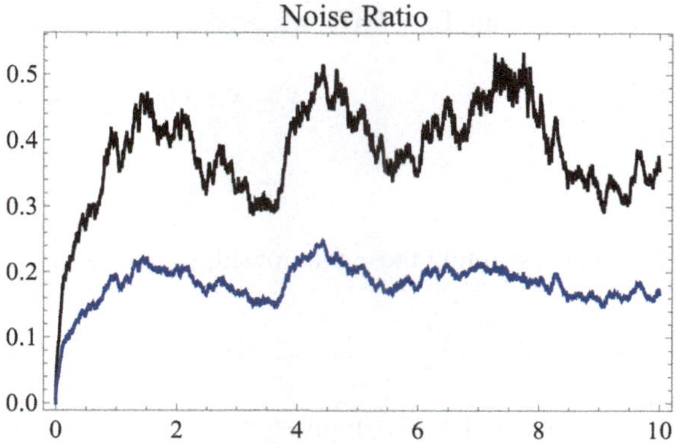

Fig. 7.2 Noise ratio for $\alpha = 0.4, 0.2$ are in black and blue, respectively

in a closed system; we denote the concentrations of the species without square bracket notation. k_f is the rate coefficient of the forward reaction and k_r is the rate coefficient of the reverse reaction. Since the system is assumed to be closed for mass transfer across the boundary, we write,

$$x + y + z = \lambda_0 \tag{7.30}$$

for any given time, and λ_0 is a constant. The initial conditions are,

$$x = x_0, y = y_0, z = z_0, \text{ and} \tag{7.31}$$

$$x_0 + y_0 + z_0 = \lambda_0. \tag{7.32}$$

This reaction has three species involved, so let us consider the deterministic case first. Assuming k_f and k_r as constant coefficients, we write the following set of ODEs:

$$\frac{dx}{dt} = -k_f xy + k_r z, \text{ and} \tag{7.34}$$

$$\frac{dz}{dt} = k_f xy - k_r z. \tag{7.35}$$

For y, it is the same equation as Eq. (7.34). By adding Eq. (3.34) and Eq. (7.35), we obtain,

$$\frac{dx}{dt} + \frac{dz}{dt} = 0, \text{ or} \tag{7.36}$$

$$\frac{dx}{dt} = -\frac{dz}{dt} = \frac{dy}{dt}. \tag{7.37}$$

By integrating Eq. (7.37), we express y and z in terms of x:

$$y = x - (x_0 - y_0), \text{ and} \tag{7.38}$$

$$z = (x_0 + z_0 - x). \tag{7.39}$$

Now substituting these relationships back to Eq. (7.34), we obtain,

$$\frac{dx}{dt} = -k_f x(x - (x_0 - y_0)) + k_r((x_0 + z_0) - x), \tag{7.40}$$

and simplifying Eq. (7.40) gives,

$$\frac{dx}{dt} = -k_f x^2 + \left(k_f(x_0 - y_0) - k_r\right)x + k_r(x_0 + z_0). \tag{7.41}$$

Equation (7.41) can be solved numerically or analytically for given initial conditions and the rate coefficients. Consider the system that has been introduced to intrinsic and extrinsic noises through irregular rate coefficients that change randomly with time and may be small numbers of molecules present in one or two species. Then we can consider all the variables, the rate coefficients except the initial conditions are stochastic variables in white noise probability space. We can model this situation using the following WSDE with the deterministic initial conditions.

$$\begin{aligned}\frac{dx(t,\omega)}{dt} &= -k_f \diamond x(t,\omega)^{\diamond 2}\\ &\quad + \left(k_f(t,\omega)(x_0 - y_0) - k_r(t,\omega)\right)\diamond x(t,\omega)\\ &\quad + k_r(t,\omega)\diamond(x_0 + z_0) + \alpha x(t,\omega)\diamond w(t,\omega)\cdot\end{aligned} \tag{7.42}$$

Here we consider α to be a given value, hence a constant. As the initial values are deterministic, we write Eq. (7.42) simplifying the notation:

$$\frac{dx}{dt} = -k_f \diamond x^{\diamond 2} + \left(k_f(x_0 - y_0) - k_r\right)\diamond x + k_r(x_0 + z_0) + \alpha x \diamond w\cdot \tag{7.43}$$

Let us write x in terms of Clarke-Ocone formula,

$x = E[x] + \int_0^t \psi(t,\omega)dW(t,\omega)$,

where $\psi(t,\omega)$ is the Malliavin derivative of x, and W is the Wiener process. Let us define

$$E[x] \equiv f(t), \text{ and} \tag{7.44}$$

$$s(t,\omega) \equiv \int_0^t \psi(t,\omega)dW(t,\omega)\cdot \tag{7.45}$$

Note f is a deterministic function. By making use of the property of Wick algebra, $E\left[x^{\diamond 2}\right] = E[x\diamond x] = (E[x])^2$, and defining $f \equiv E[x]$, we substitute, $x = f$ to Eq. (7.43) to obtain,

$$\frac{df}{dt} = -k_{f0}f^2 + \left(k_{f0}(x_0 - y_0) - k_{r0}\right)f + k_{r0}(x_0 + z_0)\cdot \tag{7.46}$$

Here we assume that k_f and k_r are positive stochastic functions as before, i.e.,

$$k_f(t,\omega) = k_{f0}\exp^{\diamond}\left[a_f w(t,\omega)\right], \text{ and} \tag{7.47}$$

$$k_r(t,\omega) = k_{r0}\exp^{\diamond}[a_r w(t,\omega)]\cdot \tag{7.48}$$

We maintain the same definitions as in Sect. 7.2 for the variables and coefficients. By substituting

$$x(t,\omega) = f(t) + s(t,\omega) \tag{7.49}$$

into Eq. (7.46) we obtain after some algebraic manipulations,

$$\begin{aligned}\frac{df}{dt} + \frac{ds}{dt} &= \left\{k_f f^2 + k_f(x_0 - y_0)f - k_r f + k_r(x_0 + z_0)\right\}\\ &+ k_f\diamond s^{\diamond 2} + \left\{(2f + (x_0 - y_0))k_f + k_r\right\}\diamond s + \alpha f w + \alpha s\diamond w\cdot\end{aligned} \tag{7.50}$$

Making use of Eq. (7.49), we get after algebraic manipulations,

$$\begin{aligned}\frac{ds}{dt} &= -(k_f - k_{f0})f^2 + \left\{(k_f - k_{f0})(x_0 - y_0) - (k_{r0} - k_r)\right\}f\\ &+ (k_r - k_{r0})(x_0 + z_0) - k_f\diamond s^{\diamond 2}\\ &+ \left\{(-2f + (x_0 - y_0))k_f - k_r\right\}\diamond s + \alpha f w + \alpha s\diamond w,\end{aligned} \tag{7.51}$$

with the initial condition, $s(t,\omega) = 0$. Using the same procedure as in Sect. 7.2, we develop a finite difference scheme based on the following difference equation:

$$\begin{aligned}s_{i+1} &= s_i - \left(k_{fi} - k_{f0}\right)f_i^2 dt + \left\{\left(k_{fi} - k_{f0}\right)(x_0 - y_0) - (k_{r0} - k_r)\right\}\\ &f_i dt + (k_{ri} - k_{r0})(x_0 + z_0)dt - k_{fi}\diamond s_i^{\diamond 2}dt\\ &+ \left\{(-2f_i + (x_0 - y_0))k_{fi} - k_{ri}\right\}\diamond s_i dt + \alpha f_i dW_i + \alpha s_i\diamond dW_i.\end{aligned} \tag{7.52}$$

We make use of Eq. (7.46) in evaluating this scheme, and in computing $s_i^{\diamond 2}dt$, we note that

$$s_i = \int_0^{t_i} \psi(t,\omega)dW(t,\omega), \tag{7.53}$$

and using the Hermite polynomials we express,

$$s_i^{\diamond 2} = \|\psi\|_i^2 h_2\left(\frac{s_i}{\|\psi\|}\right) = \|\psi\|_i^2\left(\left(\frac{s_i}{\|\psi\|_i}\right)^2 - 1\right) = s_i^2 - \|\psi\|_i^2, \tag{7.54}$$

evaluated at $t = t_i$. Here,

$$\|\psi\|_i^2 = \int \psi^2 dt. \tag{7.55}$$

Hence, we approximate,

$$\|\psi\|_i^2 dt \cong \left(\sum \psi^2 dt\right)dt = \sum(\psi^2)dt^2 \cong 0. \tag{7.56}$$

If we choose dt to be infinitesimally small, then $s_i^{\diamond 2} \cong s_i^2$. One caveat of this derivation is that Eq. (7.53) is based on ψ to be a deterministic function during the time period. Given that ψ is the expectation of the Malliavin derivative, and our assumption that it behaves like a deterministic function within a discretized minute interval of time is reasonable to keep the computations more efficient.

7.3.2 Computations

The Mathematica® program is listed in Appendix 7.2, and we compute the realizations of x, y, and z using the program as before for various values of the initial conditions of the species and the coefficients. There are many permutations of coefficients one can test mathematically but physical reality restricts the sample. The program can be used to compute 50 realizations (or more) of the species for 50 different realizations (or more) of the white noise and the Wiener increments independent to each other so the variety of scenarios can be tested. However, very high noise coefficients may not be realistic in chemical reactions in a closed system. The noise ratios for x are computed for $\alpha = \{0.1, 0.2\}$, using 50 realizations keeping all the processes at the same values, and the graphs show (not shown here) that α has significant effect upon the increase of noise. This shows that the noise needs to be considered in chemical reactions in natural settings.

7.4 Stochastic Brusselator

7.4.1 Development of the Numerical Scheme

We discuss the trimolecular reaction scheme called the Brusselator in a deterministic setting in Sect. 6.3. This scheme has a trimolecular autocatalytic step and the Brusselator has been experimentally studied by physicists as a complex example of dissipative structures. The deterministic reaction scheme given in Sect. 6.3 is repeated here:

$$A \to^{k_1} X,$$

$$B + X \to^{k_2} Y + D,$$

$$2X + Y \to^{k_3} 3X, \text{ and}$$

$$X \to^{k_4} E,$$

Where $k_i (i = 1, 2, 3, 4)$ are forward reaction rates. We assume that reverse reactions are negligible to simplify the analysis. Further we simplify the situation by assuming that D and E are removed from the system in such a way to maintain the concentrations of A, B, D, and E at constant values. Then this system reduces to a system of two variable concentrations, X and Y. We can write the following ODEs by using the notations $a \equiv [A]$, $b \equiv [B]$, which are constants, and $x \equiv [X]$, $y \equiv [Y]$:

$$\frac{dx}{dt} = k_1 a - k_2 bx + k_3 x^2 y - k_4 x, \text{ and} \tag{7.57}$$

$$\frac{dy}{dt} = k_2 bx - k_3 x^2 y \tag{7.58}$$

with $x[0] = x_0, y[0] = y_0$ as initial conditions. We also see in Sect. 6.3 that the Brusselator is unstable when

$$b > \frac{k_4}{k_2} + \frac{k_4}{k_2}\frac{k_1^2}{k_4^2} a^2. \tag{7.59}$$

We propose the following system of WSDEs for the stochastic version of the Brusselator:

$$\frac{dx(t,\omega)}{dt} = k_1(t,\omega)a - bk_2(t,\omega)\Diamond x(t,\omega)$$
$$+ k_3(t,\omega)\Diamond x(t,\omega)^2 \Diamond y(t,\omega)$$

$$- k_4(t,\omega)\Diamond x(t,\omega) + \alpha_1 x(t,\omega)\Diamond w_1(t,\omega), \text{ and} \tag{7.60}$$

$$\begin{aligned}\frac{dy(t,\omega)}{dt} &= bk_2(t,\omega)\Diamond x(t,\omega) - k_3(t,\omega)\Diamond x(t,\omega)^2\Diamond y(t,\omega)\\ &+ \alpha_2 y(t,\omega)\Diamond w_2(t,\omega),\end{aligned} \tag{7.61}$$

where α_1 and α_2 are constant coefficients for the white noise processes, w_1 and w_2, respectively. As in the previous case studies, we assume that

$$k_1(t,\omega) = k_{10}\exp^{\Diamond}\left[a_1 w_1^k(t,\omega)\right], \tag{7.62}$$

$$k_2(t,\omega) = k_{20}\exp^{\Diamond}\left[a_2 w_2^k(t,\omega)\right], \tag{7.63}$$

$$k_3(t,\omega) = k_{30}\exp^{\Diamond}\left[a_3 w_3^k(t,\omega)\right], \text{ and} \tag{7.64}$$

$$k_4(t,\omega) = k_{40} exp^{\Diamond}\left[a_4 w_4^k(t,\omega)\right], \tag{7.65}$$

and $a_i (i = 1, .., 4)$ are amplitude factors for the white noise processes, $w_i^k (i = 1, .., 4)$ are independent white noise processes. This model has total of 10 parameters to assign values giving the flexibility for computational experimentations. We develop a similar numerical scheme as before making the following assumptions,

$$x = E[x] + \int_0^t \psi_1(t,\omega)dW_1(t,\omega), \text{ and} \tag{7.66}$$

$$y = E[y] + \int_0^t \psi_2(t,\omega)dW_2(t,\omega), \tag{7.67}$$

where $\psi_1(t,\omega)$ and $\psi_2(t,\omega)$ are the Malliavin derivatives, and dW_1 and dW_2 are the independent Wiener increments. We use the notation that

$$f_1 \equiv E[x], \tag{7.68}$$

$$s_1 \equiv \int_0^t \psi_1(t,\omega)dW_2(t,\omega)\cdot \tag{7.69}$$

$$f_2 \equiv E[y], \text{ and} \tag{7.70}$$

$$s_2 \equiv \int_0^t \psi_2(t,\omega)dW_2(t,\omega). \tag{7.71}$$

Then we write the ODEs for f_1 and f_2 as

$$\frac{df_1}{dt} = k_1 a - k_2 b f_1 + k_3 f_1^2 f_2 - k_4 f_1, \text{ and} \tag{7.72}$$

$$\frac{df_2}{dt} = k_2 b f_1 - k_3 f_1^2 f_2 \tag{7.73}$$

with $f_1[0] = x_0, f_2[0] = y_0$. We have

$$x = f_1 + s_1 \text{and} \tag{7.74}$$

$$y = f_2 + s_2 \tag{7.75}$$

with $s_1(0,\omega) = s_2(0,\omega) = 0$ as initial conditions for s_1 and s_2. By substituting Eqs. (7.74), (7.75) into Eqs. (7.60), (7.61), we obtain the system of WSDEs for the model in term of f_1, f_2, s_1 and s_2, and then we develop the finite difference scheme as in the previous examples. We discuss the results next.

7.4.2 Computational Results

We develop a Mathematica® program for the numerical scheme and a large number of different combinations of coefficients (parameters) are used to test the scheme. For brevity, we only discuss a few interesting examples here. First, for the coefficients $k_{10} = 1.0$; $k_{20} = 1.0$; $k_{30} = 1.0$; $k_{40} = 1.0$; a = 1.0; b = 3.0; $a_1 = a_2 = a_3 = a_4 = 1.0$; and $\alpha_1 = \alpha_2 = 0.3$.

For these values the deterministic ODEs are unstable according to Eq. (7.59), hence we see the oscillations in Fig. 7.3 for the deterministic solutions, and the stochastic solutions show similar oscillatory patterns, but the peaks and troughs have shifted, and the peaks in stochastic case are higher than those of the deterministic case. Different realizations for the same α values depict the same tendencies.

If we change α values to $\alpha_1 = 0.6$ and $\alpha_2 = 0.4$ without any other changes, then a realization of the stochastic solutions is given in Fig. 7.4.

The increase of noise in the WSDEs would change the oscillatory behavior of the system drastically, and we see that further increase of noise would result in obliteration of the oscillatory behavior. At the same time, if we have $\alpha_1 = \alpha_2 = 0.0$, i.e., without noise in the system except the intrinsic noise in the reaction rate coefficients, we find that the oscillatory behavior is not affected by the noise amplitudes (a_i s) in the reaction coefficients significantly. We also investigate the system behaviors when

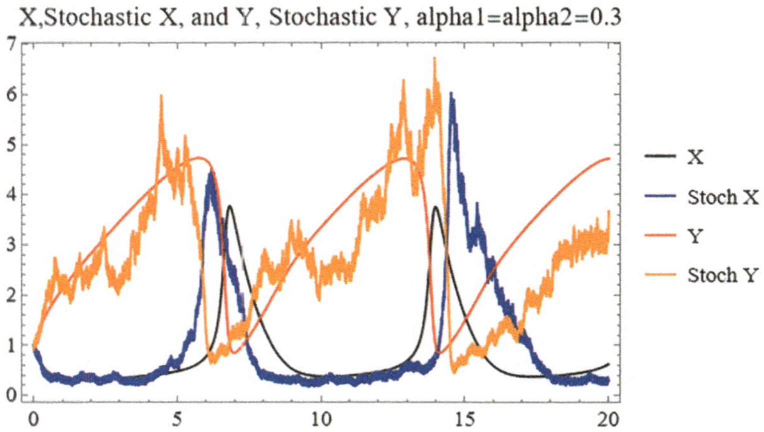

Fig. 7.3 Deterministic and stochastic solutions for x and y for $\alpha_1 = \alpha_2 = 0.3$

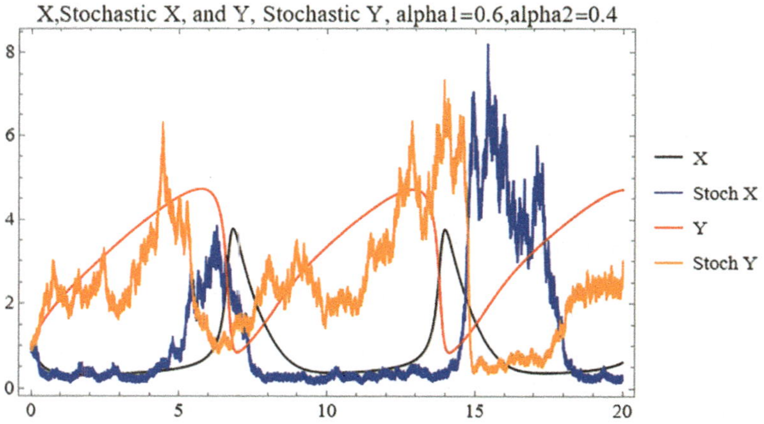

Fig. 7.4 Deterministic and stochastic solutions for x and y for $\alpha_1 = 0.6, \alpha_2 = 0.4$

the Brusselator system is stable by changing the values so that, $b \leq \frac{k_4}{k_2} + \frac{k_4}{k_2}\frac{k_1^2}{k_4^2}a^2$. By using the coefficients $k_{10} = 1.0$; $k_{20} = 1.0$; $k_{30} = 1.0$; $k_{40} = 1.0$; $a = 1.0$; $b = 1.5$; $a_1 = a_2 = a_3 = a_4 = 1.0$; and $\alpha_1 = \alpha_2 = 0.1$, we compute the solutions and two realizations for stochastic x and y are given in Fig. 7.5.

The stochastic solutions are fluctuating around the stable solutions, and we note that the stochastic solutions are more sensitive to changes in α_1 and α_2 than when the system is unstable.

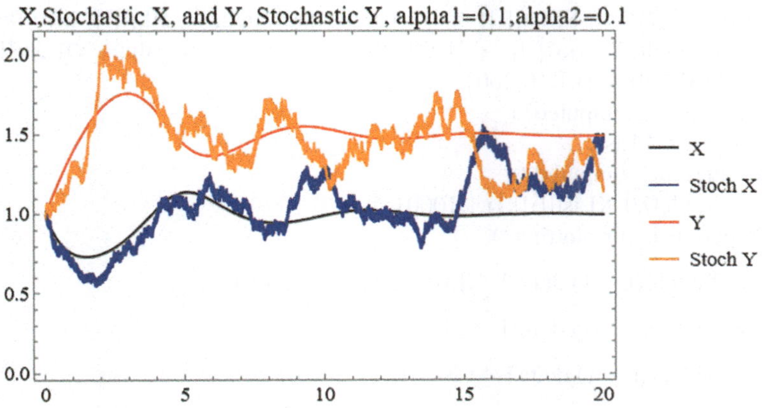

Fig. 7.5 A stable Brusselator behavior and its stochastic counterparts

Appendix 7.1: Mathematica® Code for $X \rightleftharpoons Y$

```
(*Chapter 7_1 notebook for x ⇌ y reaction; numerical solutions*).
   (*First, we assign values to the coefficients*).
   kf0 = 0.5; af = 3.0; alpha = 0.20; kr0 = 0.20; ar = 3.0; deltaT = 0.001;
   X0 = 1.0; Y0 = 0.3; λ0 = X0 + Y0;
   (* We use WhiteNoiseProcess in Mathematica to generate the white noise with
square root of deltaT as the argument*).
   WdataList =
      Normal[RandomFunction[WhiteNoiseProcess[0.0316228],
      {0, 10,001}]];
   (*Stochastic functions for rate coefficients are computed*).

kfp = Table[ Exp[af WdataList[[1, i, 2]]], {i, 1, 10001}];

kf = kf0 kfp;

krp = Table[ Exp[ar WdataList[[1, i, 2]]], {i, 1, 10001}];

kr = kr0 krp;
   (*Wiener increments are computed *).
   dBList=Table[N[Random[NormalDistribution[0.0,Sqrt[deltaT]]], 32], {10000}];
   (*Compute function f *).
   fList = Table[ ((kr0 λ0)/(kf0 + kr0)) + (X0 -((kr0 λ0)/(kf0 + kr0))) Exp[-(kf0 +
kr0)deltaT(i-1)],{i,1,10,001}];
   (*Initialize s function to zero*).
   SList=Table[0.0,{i,1,10001}];
   (*Compute s values *).
   Do[
```

```
SList[[i + 1]] = SList[[i]] + (kr[[i]]-kr0)λ0 deltaT -((kf[[i]]-kf0)-(kr[[i]]-kr0))fList[[i]]deltaT -(kf[[i]] + kr[[i]])SList[[i]]deltaT + alpha (fList[[i]] + SList[[i]])dBList[[i]],{i,1,10,000}];
   (*x values are computed*).
   XList=fList+SList;
   (*y values are computed*).
   YList=Table[λ0-XList[[i]],{i,1,10001}];
   (*Prepare lists for plotting*).
XtList = Table[{(i − 1) deltaT, XList[[i]]}, {i, 1, 10001}];
YtList = Table[{(i − 1) deltaT, YList[[i]]}, {i, 1, 10001}];
StList = Table[{(i − 1) deltaT, SList[[i]]}, {i, 1, 10001}];
ftList = Table[{(i − 1) deltaT, fList[[i]]}, {i, 1, 10001}];
ListLinePlot[{XtList,YtList,StList, ftList},Frame->True,
Axes->True, Joined->True, PlotLabel->"x, y, s, and f",
AxesLabel->{"t"},PlotStyle->{Blue, Red,Black, Green,Thick},
PlotLegends->Placed[{"x","y","s","f"},{Center,Below}],
LabelStyle->{FontFamily->"Times New Roman",14,GrayLevel[0]}]
SList=Table[0.0,{i,1,10001}];
   (*Compute s values *)
   Do[
SList[[i + 1]] = SList[[i]] + (kr[[i]]-kr0)λ0 deltaT -((kf[[i]]-kf0)-(kr[[i]]-kr0))fList[[i]]deltaT -(kf[[i]] + kr[[i]])SList[[i]]deltaT + alpha (fList[[i]] + SList[[i]])dBList[[i]],{i,1,10,000}];
   (*x values are computed*)
XList=fList+SList;
(*y values are computed*)
YList=Table[λ0-XList[[i]],{i,1,10001}];
(*Prepare lists for plotting*)
XtList = Table[{(i − 1) deltaT, XList[[i]]}, {i, 1, 10001}];
YtList = Table[{(i − 1) deltaT, YList[[i]]}, {i, 1, 10001}];
StList = Table[{(i − 1) deltaT, SList[[i]]}, {i, 1, 10001}];
ftList = Table[{(i − 1) deltaT, fList[[i]]}, {i, 1, 10001}];
ListLinePlot[{XtList,YtList,StList, ftList}, Frame- > True,
Axes- > True, Joined- > True, PlotLabel- > "x, y, s, and f",
```

```
AxesLabel- > {"t"}, PlotStyle- > {Blue, Red,Black, Green,Thick},

PlotLegends- > Placed[{"x","y","s","f"},{Center, Below}],LabelStyle- >
{FontFamily- > "Times New Roman",14,GrayLevel[0]}]
   (*We will do multiple realizations with n = 50; We will calculate noise ratio*).
   WdataList = Normal[RandomFunction[WhiteNoiseProcess[0.0316228], {0,
10,001},50]];

kfp=Table[ Exp[af WdataList[[j,i,2]]],{i,1,10001},{j,1,50}];

kfpT=Transpose[kfp];

kf=kf0 kfpT;

krp=Table[ Exp[ar WdataList[[j,i,2]]],{i,1,10001},{j,1,50}];

krpT=Transpose[krp];

kr=kr0 krpT;

dBList=Table[N[Random[NormalDistribution[0.0, Sqrt[deltaT]]], 32],
{10000},50];
   dBListT=Transpose[dBList];
   fList=Table[ ((kr0 λ0)/(kf0 + kr0)) + (X0 -((kr0 λ0)/(kf0 + kr0))) Exp[-
(kf0+kr0)deltaT(i-1)],{i,1,10001},50];
   fListT=Transpose[fList];
   SList=Table[0.0,{i,1,10001},50];
   SListT=Transpose[SList];
   (*Compute s function realization*)

Do[

SListT[[j,i + 1]] = SListT[[j,i]] + (kr[[j,i]]-kr0)λ0 deltaT -((kf[[j,i]]-kf0)-(kr[[j,i]]-
kr0))fListT[[j,i]]deltaT -(kf[[j,i]] + kr[[j,i]])SListT[[j,i]]deltaT + alpha (fListT[[j,i]]
+ SListT[[j,i]])dBListT[[j,i]],{i,1,10,000},{j,1,50}];

XListT=fListT+SListT;

YListT=Transpose[Table[λ0-XListT[[j,i]],{i,1,10001},{j,1,50}]];
   meanXListT=Mean[XListT];
   varXListT=Variance[XListT];
   noiseRatio =
   Table[(Sqrt[varXListT[[i]]]/meanXListT[[i]]),{i,1,10,001}];
   noiseRatioT = Table[{(i − 1) deltaT, noiseRatio[[i]]}, {i, 1, 10001}];
   ListLinePlot[{noiseRatioT},Frame->True,Axes->True,
   Joined->True, PlotLabel->"Noise Ratio for alpha=0.2",
   AxesLabel->{"t"},PlotStyle->{Blue,Thick},
   LabelStyle->{FontFamily->"Times New Roman",14,GrayLevel[0]}]
   (* We change alpha = 0.4; other coefficients are the same*)
   alpha = 0.40;
   (*Recompute s *)
```

```
SList=Table[0.0,{i,1,10001},50];
SListT=Transpose[SList];

Do[

SListT[[j,i + 1]] = SListT[[j,i]] + (kr[[j,i]]-kr0)λ0 deltaT -((kf[[j,i]]-kf0)-(kr[[j,i]]-kr0))fListT[[j,i]]deltaT -(kf[[j,i]] + kr[[j,i]])SListT[[j,i]]deltaT + alpha (fListT[[j,i]] + SListT[[j,i]])dBListT[[j,i]],{i,1,10,000},{j,1,50}];
XListT=fListT+SListT;
YListT=Transpose[Table[λ0-XListT[[j,i]],{i,1,10001},{j,1,50}]];
meanXListT=Mean[XListT];
varXListT=Variance[XListT];
noiseRatio=Table[(Sqrt[varXListT[[i]]]/meanXListT[[i]]),{i,1,10001}];
noiseRatioT = Table[{(i – 1) deltaT, noiseRatio[[i]]}, {i, 1, 10001}];

ListLinePlot[{noiseRatioT}, Frame- > True, Axes- > True,

Joined- > True,

PlotLabel- > "Noise Ratio ", AxesLabel- > {"t"},

PlotStyle- > {Black,Thick},

LabelStyle- > {FontFamily- > "Times New Roman",14, GrayLevel[0]}]
(*Show both noise ratio curve in the same plot. "**" indicates the output number of each plot*).
Show[%**,%**, PlotRange->All]
```

Appendix 7.2: Mathematica® Code for $X + Y \rightleftharpoons Z$

```
(*Chapter 7_2 notebook for x + y ⇌ z reaction; numerical solutions*)
(*First, we assign values to the coefficients*)

kf0=0.5; af=3.0; alpha=0.20;

kr0=0.20; ar=3.0;deltaT=0.001;

X0=1.0; Y0=0.6;Z0=0.3;
(* We use WhiteNoiseProcess in Mathematica to generate the white noise with square root of deltaT as the argument*)
WdataList = Normal[RandomFunction[WhiteNoiseProcess[0.0316228],{0, 10,001}]];
(*Stochastic function for rate coefficients is computed*)
kfp=Table[ Exp[af WdataList[[1,i,2]]],{i,1,10001}];
kf=kf0 kfp;
(*Plot to check*)

ListLinePlot[kf, Frame->True, Axes->True, PlotLabel->"kf",
```

```
AxesLabel->{"t"},PlotStyle->{Black,Thin},PlotRange->All,
LabelStyle->{FontFamily->"Times New Roman",14,GrayLevel[0]}]
krp=Table[ Exp[ar WdataList[[1,i,2]]],{i,1,10001}];
kr=kr0 krp;
ListLinePlot[kr, Frame->True, Axes->True, PlotLabel->"kr",
AxesLabel->{"t"}, PlotStyle->{Black,Thin}, PlotRange->All,
LabelStyle->{FontFamily->"Times New Roman",14, GrayLevel[0]}]
   (*Wiener increments are computed *)
   dBList = Table[N[Random[NormalDistribution[0.0, Sqrt[deltaT]]], 32],
{10,000}];
ListLinePlot[dBList,Frame- > True, Axes- > True, PlotLabel- > "deltaW",
AxesLabel- > {"t"}, PlotStyle- > {Black,Thin}, PlotRange- > All,
LabelStyle- > {FontFamily- > "Times New Roman",14, GrayLevel[0]}]
   (*Compute function f numerically*).
   fsol = NDSolve[{f'[t] + kf0 f[t]^2 -(kf0(X0-Y0)-kr0)f[t]-kr0 (X0 + Z0) = =
0.0,f[0] = = X0},f,{t,0,12}].
Plot[Evaluate[f[t] /. fsol],{t,0,10}, Frame- > True, Axes- > True, PlotLabel- > "f =
E[x]", AxesLabel- > {"t"},
PlotStyle- > {Black,Thick}, PlotRange- > All,
PlotLegends->Placed[{"x"},{Center, Below}],
LabelStyle->{FontFamily->"Times New Roman",14, GrayLevel[0]}]
   fList=Table[0.0,{i,1,10001}];
   Do[ t = deltaT(i-1); f = Evaluate[f[t] /.fsol]; fList[[i]] = f, {i,1,10,001}];
   fList=Flatten[fList];
   ftList = Table[{(i - 1) deltaT, fList[[i]]}, {i, 1, 10001}];
ListLinePlot[ftList,Frame- > True,Axes- > True, PlotLabel- > "f = E[x]",
AxesLabel- > {"t"},PlotStyle- > {Black,Thick},PlotRange- > All,
LabelStyle- > {FontFamily- > "Times New Roman",14, GrayLevel[0]}]
   (*Initialize s function to zero*)
   SList=Table[0.0,{i,1,10001}];
   (*Compute s values *).
Do[SList[[i+1]]=SList[[i]]- (kf[[i]]-kf0)*((fList[[i]])^2)*deltaT
+ ((kf[[i]]-kf0)*(X0-Y0)-(kr0-kr[[i]]))*fList[[i]]*deltaT
+ (kr[[i]]-kr0)*(X0 + Z0)*deltaT + kf[[i]] *(SList[[i]]^2)* deltaT
+ (-2*fList[[i]]*kf[[i]] + (X0-Y0)*kf[[i]]-kr[[i]])*SList[[i]]*deltaT
```

```
+ alpha *(fList[[i]] + SList[[i]])*dBList[[i]],{i,1,10,000}]
   ListLinePlot[SList, Frame->True, Axes->True, PlotLabel->"S",
   AxesLabel->{"t"},PlotStyle->{Black,Thick},PlotRange->All,
   LabelStyle->{FontFamily->"Times New Roman",14, GrayLevel[0]}]
   (*x values are computed*)
   XList=fList+SList;
   (*y and z values are computed*)
   YList=Table[(XList[[i]]-(X0-Y0)),{i,1,10001}];
   ZList=Table[(-XList[[i]]+(X0+Z0)),{i,1,10001}];
   (*Prepare lists for plotting*)

XtList = Table[{(i – 1) deltaT, XList[[i]]}, {i, 1, 10,001}];

YtList = Table[{(i – 1) deltaT, YList[[i]]}, {i, 1, 10001}];

ZtList = Table[{(i – 1) deltaT, ZList[[i]]}, {i, 1, 10001}];

StList = Table[{(i – 1) deltaT, SList[[i]]}, {i, 1, 10001}];

ftList = Table[{(i – 1) deltaT, fList[[i]]}, {i, 1, 10001}];
   ListLinePlot[{XtList,YtList,ZtList,StList, ftList},Frame->True,
   Axes->True,Joined->True, PlotLabel->"x,y,z,s, and f", AxesLabel->{"t"},
   PlotStyle->{Blue, Red,Black, Green,Purple,Thick},
   PlotLegends->Placed[{ 'x","y","z","s","f"},{Center, Below}],
   LabelStyle->{FontFamily->"Times New Roman",14, GrayLevel[0]}]
   (*We will do multiple realizations with n = 50; We will calculate noise ratio*).

Clear[f,t]

WdataList = Normal[RandomFunction[WhiteNoiseProcess[0.0316228], {0,
10,001},50]];

kfp=Table[ Exp[af WdataList[[j,i,2]]],{i,1,10001},{j,1,50}];

kfpT=Transpose[kfp];

kf=kf0 kfpT;

krp=Table[ Exp[ar WdataList[[j,i,2]]],{i,1,10001},{j,1,50}];

krpT=Transpose[krp];

kr=kr0 krpT;
   dBList = Table[N[Random[NormalDistribution[0.0, Sqrt[deltaT]]], 32],
{10,000},50];
   dBListT=Transpose[dBList];
   fList=Table[0.0,{i,1,10001}];
   Clear[f,t]
   Do[ t=deltaT(i-1);f=Evaluate[f[t] /.fsol]; fList[[i]]=f, {i,1,10001}];
   fListT=Flatten[fList];
   SList=Table[0.0,{i,1,10001},50];
```

```
SListT=Transpose[SList];
(*Compute s function realization*)
Do[SListT[[j,i+1]]=SListT[[j,i]]- (kf[[j,i]]-kf0)*((fListT[[i]])^2)*deltaT
+ ((kf[[j,i]]-kf0)*(X0-Y0)-(kr0-kr[[j,i]]))*fListT[[i]]*deltaT
+ (kr[[j,i]]-kr0)*(X0 + Z0)*deltaT + kf[[j,i]] *(SListT[[j,i]]^2)* deltaT
+ (-2*fListT[[i]]*kf[[j,i]] + (X0-Y0)*kf[[j,i]]-kr[[j,i]])*SListT[[j,i]]*deltaT
+ alpha *(fListT[[i]] + SListT[[j,i]])*dBListT[[j,i]],{i,1,10,000},{j,1,50}];
ListLinePlot[SListT, Frame- > True, Axes- > True, Joined- > True,
PlotLabel- > "S realizations for alpha = 0.2", AxesLabel- > {"t"},
PlotStyle- > Automatic, LabelStyle- > {FontFamily- > "Times New Roman",14,GrayLevel[0]}]
varSListT=Variance[SListT];
noiseRatio=Table[(Sqrt[varSListT[[i]]]/fListT[[i]]),{i,1,10001}];
noiseRatioT = Table[{(i – 1) deltaT, noiseRatio[[i]]}, {i, 1, 10001}];
ListLinePlot[{noiseRatioT}, Frame->True, Axes->True, Joined->True,
PlotLabel->"Noise Ratio for alpha=0.2", AxesLabel->{"t"},
PlotStyle->{Blue,Thick}, LabelStyle->{FontFamily->"Times New Roman",14,GrayLevel[0]}]
(* We change alpha = 0.1; other coefficients are the same*)
alpha = 0.10;
(*Recompute s *)
SList = Table[0.0,{i,1,10,001},50];
SListT = Transpose[SList];
Do[SListT[[j,i + 1]] = SListT[[j,i]]- (kf[[j,i]]-kf0)*((fListT[[i]])^2)*deltaT
+ ((kf[[j,i]]-kf0)*(X0-Y0)-(kr0-kr[[j,i]]))*fListT[[i]]*deltaT
+ (kr[[j,i]]-kr0)*(X0 + Z0)*deltaT + kf[[j,i]] *(SListT[[j,i]]^2)* deltaT
+ (-2*fListT[[i]]*kf[[j,i]] + (X0-Y0)*kf[[j,i]]-kr[[j,i]])*SListT[[j,i]]*deltaT
+ alpha *(fListT[[i]] + SListT[[j,i]])*dBListT[[j,i]],{i,1,10,000},{j,1,50}];
varSListT=Variance[SListT];
noiseRatio=Table[(Sqrt[varSListT[[i]]]/fListT[[i]]),{i,1,10001}];
noiseRatioT = Table[{(i – 1) deltaT, noiseRatio[[i]]}, {i, 1, 10001}];
ListLinePlot[{noiseRatioT},Frame- > True,Axes- > True,
Joined- > True,
PlotLabel- > "Noise Ratio ", AxesLabel- > {"t"},
PlotStyle- > {Black,Thick},
LabelStyle- > {FontFamily- > "Times New Roman",14,GrayLevel[0]}]
(*Show both noise ratio curves in the same plot. "**" indicates the output number of each plot*)
Show[%**,%**, PlotRange->All]
```

Chapter 8
Stochasticity in Chiral Symmetry Breaking

8.1 Introduction

Most chemical transformations in nature are based on irreversible chemical reactions. Irreversibility often leads to dismantling of order in a system increasing entropy near thermodynamic equilibrium but gives rise to order far from equilibrium (Prigogine & Stengers, 2017), and the self-organized behaviors in open systems far from equilibrium are called dissipative structures, which occur when the environment interacts with species in the system. Often these interactions cause fluctuations in the system -they are often called "noise"- in turn break down the deterministic behavior of chemical reactions. As we have seen in previous chapters, the deterministic ODEs that describe the chemical reactions often become unstable under certain conditions. If we can identify the parameters that influence the stability of a system, by changing the parameters thus identified we could drive the system toward a bifurcation point. Near a bifurcation point, fluctuations cause the system to move along different branches of the bifurcation plot (Kondepudi & Prigogine, 2014). Along these branches order would be established, or in other words, fluctuations establish order in the system. As often happens, deterministic ODEs can be a guide prior to bifurcation point and the stochastic or probabilistic behavior takes over. Most often chemical master equations would provide a valid description of the system behavior, but our intention in this chapter is to show that we can still use the structure of ODEs to investigate the stochastic behavior by developing WSDE models near bifurcation point to understand the stochasticity. WSDEs are based on white noise probability space as mentioned in Chap. 2, and this space provides a more appropriate space to develop stochastic models of chemical reactions.

D. Kulasiri, *Stochastic Differential Equations for Chemical Transformations in White Noise Probability Space*, https://doi.org/10.1007/978-981-97-9392-1_8

We review the mathematical aspects of Kondepudi and Nelson model (KNM) (Kondepudi & Nelson, 1983, 1984b) on chiral symmetry breaking first and then based on the deterministic ODEs we develop WSDEs to explore stochasticity in KNM later. The emphasis here is mathematical modeling of the system and for physics of the problem we refer the reader to the references in this chapter.

8.2 Chirality and Symmetry Breaking

Chirality is a property of an object that depicts its "handedness". (The word chirality is from the Greek "cheir" meaning hand). A chiral object does not have an identical mirror image, i.e., an axis cannot be found to project one part of the object to be similar to the rest of the object. We say chiral objects have a "handedness", and this is a very important property in biomolecules. It is noted that the most key molecules has the same handedness. Kondepudi and Prigogine (2014) remark that this is quite remarkable because chemical reactions do not favor one mirror image form to the other except for "very small differences due to parity-non-conserving electroweak interactions". "Enantiomers" are mirror image structures of a chiral molecule, and L- and D- enantiomers—L for "levo" and D for "dextro"- is the convention we use here to identify the left-handed and right-handed enantiomers, respectively. If all reactions in life happen in thermodynamic equilibrium, we would not see chirality in biomolecules because chemical reactions would produce achiral molecules, i.e., no preference to either L- or D- enantiomers. Therefore, the cause for chirality is due to the chemical reactions occurring in far from thermodynamic equilibrium, in open systems subjected to fluctuations, i.e., "noise". To understand this phenomenon, Kondepudi and Nelson developed the physics and the models in a series of papers (Kondepudi & Nelson, 1984a, 1984b, 1985). We refer readers to these references for further study. We briefly discuss KNM model here to understand the dynamics before venturing into developing a stochastic model.

8.3 Kondepudi and Nelson Model (KNM)

8.3.1 Reaction Scheme

KNM is a deterministic model of autocatalytic reactions of chiral molecules. We discuss this model incorporating the recent mathematical analysis of KNM (Colwell, 2022). A scheme of chemical reactions should have catalytic activities favoring one type of enantiomer over the other. KNM is based on the following set of reactions.

$$\text{Reaction 1(R1)} : S + T \rightleftharpoons X_L \tag{8.1}$$

(k_{1f} = forward reaction rate coefficient, and k_{1r} = reverse reaction rate coefficient);

$$\text{R2} : S + T + X_L \rightleftharpoons 2X_L \left(k_{2f} \text{ and } k_{2r}\right); \tag{8.2}$$

$$\text{R3} : S + T \rightleftharpoons X_D \left(k_{1f} \text{ and } k_{1r}\right); \tag{8.3}$$

$$\text{R4} : S + T + X_D \rightleftharpoons 2X_D \left(k_{2f} \text{ and } k_{2r}\right); \tag{8.4}$$

and

$$\text{R5} : X_L + X_D \rightleftharpoons P \quad \left(k_{3f} \text{ and } k_{3r}. \text{ Assume } k_{3r} \cong 0, \text{ and } k_{3f} = k_3\right). \tag{8.5}$$

Here S and T are achiral molecules, which do not possess chirality, and R1 and R3 produce levo X (X_L) and dextro X (X_D) with same rate coefficients in both forward and reverse directions due to symmetry. We assume local thermodynamic equilibrium (see Chap. 1). The autocatalytic reactions are R2 and R4 with the same rates. In R5, X_L and X_D produce an inactive dimer P. If the system is in thermodynamic equilibrium, $[X_L] = [X_D]$, but KNM operates in an open system in which S and T are pumped into and P is pumped out in such a way that $[S]$ and $[T]$ are maintained at a fixed level, and in R5, the reverse reaction rate coefficient can be ignored based on computer simulations that show that conclusions are not significantly affected by this assumption.

8.3.2 Deterministic ODEs

Based on the assumptions mentioned above, we write the coupled ODEs as follows:

$$\begin{aligned}\frac{d[X_L]}{dx} &= k_{1f}[S][T] - k_{1r}[X_L] + k_{2f}[X_L][S][T] \\ &\quad - k_{2r}[X_L]^2 - k_3[X_L][X_D]\end{aligned} \tag{8.6}$$

and

$$\begin{aligned}\frac{d[X_D]}{dx} &= k_{1f}[S][T] - k_{1r}[X_D] + k_{2f}[X_D][S][T] \\ &\quad - k_{2r}[X_D]^2 - k_3[X_L][X_D]\end{aligned} \tag{8.7}$$

The rate coefficients are related to each other by,

$$\frac{k_{1f}}{k_{1r}} = \frac{k_{2f}}{k_{2r}} \tag{8.8}$$

due to the fact that the direct reaction equilibrium coefficients and the reverse reaction equilibrium coefficients are the same (Kondepudi & Prigogine, 2014). We define the following variables to distinguish between the asymmetric and symmetric states:

$$\alpha = \frac{X_L - X_D}{2} \tag{8.9}$$

$$\beta = \frac{X_L + X_D}{2} \tag{8.10}$$

and

$$\lambda = [S][T]. \tag{8.11}$$

When the system is in asymmetric state, $\alpha \neq 0$ and for symmetric state, $\alpha = 0$, and $\alpha, \beta, \lambda \in \mathbb{R}$ with $\lambda \geq 0$ and $\beta \geq |\alpha|$. Using these definitions,

$$[X_L] = 2\alpha + [X_D], \tag{8.12}$$

$$[X_D] = 2\beta - [X_L] \tag{8.13}$$

or

$$[X_L] = 2\beta - [X_D]. \tag{8.14}$$

In addition,

$$[X_L] = \alpha + \beta \tag{8.15}$$

and

$$[X_D] = \beta - \alpha. \tag{8.16}$$

Now substituting these relationships back to Eq. (8.6) and Eq. (8.7), the following ODEs are obtained:

$$\frac{d\alpha}{dt} = -k_{1r}\alpha + k_{2f}\lambda\alpha - 2k_r\alpha\beta \tag{8.17}$$

and

$$\frac{d\beta}{dx} = k_{1f}\lambda - k_{1r}\beta - k_{2f}\lambda\beta - k_{2r}(\beta^2 + \alpha^2) - k_3(\beta^2 - \alpha^2). \tag{8.18}$$

The initial conditions for α and β should be chosen in such a way so that they meet the asymmetric conditions, which we discuss later in the chapter. We explore the asymmetric steady states next.

8.3.3 Steady States When $\alpha \neq 0$

We find the asymmetric steady states by making

$$\frac{d\alpha}{dt} = \frac{d\beta}{dt} = 0 \tag{8.19}$$

in Eqs. (8.17) and (8.18), respectively. By solving the resulting equations, we obtain the asymmetric steady state for β (β_a),

$$\beta_a = \frac{k_{2f}\lambda - k_{1r}}{2k_{2r}} \tag{8.20}$$

and

$$\alpha = \pm\sqrt{\beta_a^2 - \frac{k_{1f}\lambda}{k_3 - k_{2r}}}. \tag{8.21}$$

For $\beta \geq |\alpha|$,

$$\frac{k_{1f}\lambda}{k_3 - k_{2r}} \geq 0, \tag{8.22}$$

and for this to happen for a physically realizable situation ($\lambda > 0$),

$$k_3 > k_{2r}. \tag{8.23}$$

This is a necessary condition for the existence of asymmetric steady states. (See Colwell (2022) for further details.) Denoting the positive asymmetric steady state value of α by α_a,

$$\alpha_a = \sqrt{\beta_a^2 - \frac{k_{1f}\lambda}{k_3 - k_{2r}}}. \tag{8.24}$$

To find λ values for which $(\alpha, \beta) = (\pm\alpha_a, \beta_a)$ possible, we note that α_a is real and non-zero, if and only if,

$$\beta_a^2 > \frac{k_{1f}\lambda}{k_3 - k_{2r}} \tag{8.25}$$

but from Eq. (8.20),

$$\beta_a = \frac{k_{2f}\lambda - k_{1r}}{2k_{2r}}. \tag{8.26}$$

Then,

$$\left(\frac{k_{2f}\lambda - k_{1r}}{2k_{2r}}\right)^2 > \frac{k_{1f}\lambda}{k_3 - k_{2r}}, \tag{8.27}$$

only if

$$k_{2f}\lambda - k_{1r} > 0, \tag{8.28}$$

otherwise λ values do not exist for which $(\alpha, \beta) = (\pm\alpha_a, \beta_a)$. Assuming $k_{2f}\lambda - k_{1r} > 0$ and $k_3 > k_{2r}$, we state Eq. (8.27) as,

$$k_{2f}^2\lambda^2 - s\lambda + k_{1r}^2 > 0, \tag{8.29}$$

where

$$s = 2k_{2f}k_{1r} + \frac{4k_{2r}^2 k_{1f}}{k_3 - k_{2r}} > 2k_{2f}k_{1r}. \tag{8.30}$$

The inequality in Eq. (8.30) is due to Eq. (8.29).

$$k_{2f}^2\lambda^2 - s\lambda + k_{1r}^2 = 0, \tag{8.31}$$

has the following solutions:

$$\lambda_0 = \frac{s - \sqrt{s^2 - 4k_{2f}^2 k_{1r}^2}}{2k_{2f}^2} \tag{8.32}$$

and

$$\lambda_c = \frac{s + \sqrt{s^2 - 4k_{2f}^2 k_{1r}^2}}{2k_{2f}^2} \tag{8.33}$$

As $\lambda \to \pm\infty$, $k_{2f}^2\lambda^2 - s\lambda + k_{1r}^2 \to \infty$,

then the inequality in Eq. (8.27) is true if $\lambda < \lambda_0$ and $\lambda > \lambda_c$. Let us look at these two ranges of λ. The first inequality can be expressed by using Eq. (8.29),

$$2k_{2f}\lambda < \frac{s}{k_{2f}} - \sqrt{\left(\frac{s}{k_{2f}}\right)^2 - (2k_{1r})^2} \tag{8.34}$$

This leads to by the triangle of inequality $2k_{2f}\lambda < 2k_{1r}$, and this condition contradicts Eq. (8.28), $k_{2f}\lambda - k_{1r} > 0$. Therefore, when $\lambda < \lambda_0$, asymmetric steady states cannot exist. But for $\lambda > \lambda_c$,

$$2k_{2f}\lambda > \frac{s}{k_{2f}} + \sqrt{\left(\frac{s}{k_{2f}}\right)^2 - (2k_{1r})^2} \quad \text{and} \tag{8.35}$$

this is true when $k_{2f}\lambda - k_{1r} > 0$. Therefore, for asymmetric steady states to exist, $\lambda > \lambda_c$, and λ_c is the critical value for λ for asymmetric steady states to exist. To summarize, for asymmetric steady states to exist $k_3 > k_{2r}$ and $\lambda > \lambda_c$. We use these conditions to check the coefficients of the system whether they would lead to asymmetric steady states.

8.3.4 Steady States When $\alpha = 0$

Symmetric steady states occur when $\alpha = 0$, and $\frac{d\beta}{dt} = 0$, i.e.,

$$k_{1f}\lambda - k_{1r}\beta - k_{2f}\lambda\beta - k_{2r}\left(\beta^2 + \alpha^2\right) - k_3\left(\beta^2 - \alpha^2\right) = 0. \tag{8.36}$$

The solution to Eq. (8.36) is,

$$\beta = \frac{2k_{2r}\beta_a \pm \sqrt{(2k_{2r}\beta_a)^2 + 4(k_{2r} + k_3)k_{1f}\lambda}}{2(k_{2r} + k_3)}, \tag{8.37}$$

and for $\beta > 0$, then the symmetric β value is,

$$\beta_s = \frac{2k_{2r}\beta_a + \sqrt{(2k_{2r}\beta_a)^2 + 4(k_{2r} + k_3)k_{1f}\lambda}}{2(k_{2r} + k_3)}, \tag{8.38}$$

and this can occur for any reaction rate coefficients and any λ values when $\alpha = 0$. Because we are only interested in existence in chirality, we explore only the asymmetric steady states.

8.3.5 Bifurcation Points

The bifurcation points are the values of λ where the asymmetric steady states and symmetric steady states coincide, i.e., $\alpha_a = \alpha_s = 0$ and $\beta_a = \beta_s$. By inserting Eq. (8.24) for β_a in Eq. (8.36), and solving for β_a, we obtain,

$$\beta_a^2 = \frac{k_{1f}\lambda}{k_3 - k_{2r}}.$$

This is where λ begins to be greater than λ_c. Therefore, λ_c is a possible bifurcation point. And it can be shown that $\beta_a > \beta_s$ as λ increases beyond λ_c (Colwell, 2022).

8.3.6 Stability of the Asymmetric Steady States

Since we are only interested in the asymmetric steady states in this chapter, we attend only the stability of asymmetric steady states using linear stability analysis, which is based on the Jacobian of the differential operator. It can be shown that the real parts of eigenvalues of Jacobian matrix constructed using Eq. (8.17) and Eq. (8.18) are negative under the conditions for the existence of asymmetric steady states, hence the steady states are stable when $\lambda > \lambda_c$. (See Colwell (2022) for mathematical analysis).

Following Colwell (2022), we summarize the necessary and sufficient conditions for chirality to exist:

- The system has a single symmetric steady state at $(0, \beta_s)$ for any value of λ;
- If $k_3 > k_{2r}$ is not true, asymmetric steady states do not exist; and
- If $k_3 > k_{2r}$ is true, then λ_c exists such that:

  1. Asymmetric steady states exist exactly when $\lambda > \lambda_c$;
  2. If $\lambda > \lambda_c$, there are exactly two asymmetric steady states $(\pm\alpha_a, \beta_a)$, which are stable whenever they exist;
  3. The symmetric steady states are stable if $\lambda < \lambda_c$ and unstable if $\lambda > \lambda_c$;
  4. All three steady states $(0, \beta_s)$ and $(\pm\alpha_a, \beta_a)$ coincide if $\lambda > \lambda_c$; and
  5. $k_3 > k_{2r}$ is both necessary and sufficient condition for bifurcation.

8.4 Computations Based on KNM

As KNM is a deterministic model, we briefly discuss the numerical solutions following Kondepudi and Prigogine (2014) based on the Mathematica® program given in Appendix. As discussed in the previous sections, the coefficients of the model need to be carefully chosen to see the chirality of the reactions. We use the following values for the coefficients as an example: $k1f = 0.5$; $k_{1r} = 0.2$; $k_{2f} = 0.5$;

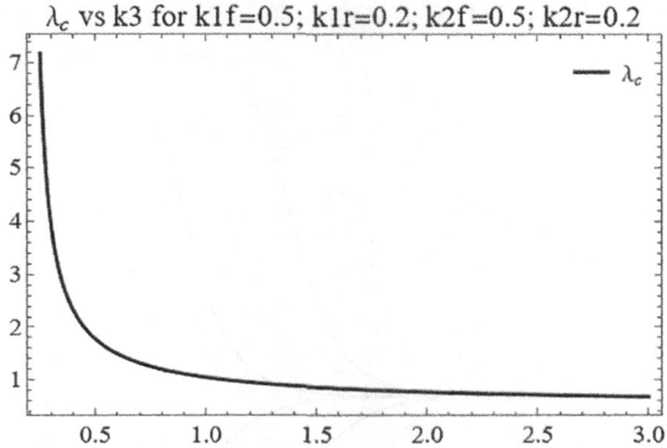

Fig. 8.1 Critical value λ_c (vertical axis) versus k_3 horizonal axis) for $k_{1\mathrm{f}} = 0.5$; $k_{1\mathrm{r}} = 0.2$; and $k_{2\mathrm{f}} = 0.5$

$k_{3\mathrm{f}} = k_3 = 1.5$; $k_{3\mathrm{r}} = 10^{-3}$; $[S] = 1.25$; $[T] = [S]$; and $[P] = 0.5$. These values give $\lambda = [T][S] = 1.5625$ and note $k_3 > k_{2r}$ which satisfies the necessary and sufficient condition for bifurcation. We use a relatively negligible value for $k_{3\mathrm{r}}$ to facilitate the computations of entropy production rates later. Based on the equations given in Sect. 8.3.2, we develop a module to compute the critical values for λ, λ_c, as a function of the coefficients. Keeping all the coefficients at the given values except $k_{3\mathrm{f}}$ and making $k_{3\mathrm{r}} = 0$, we plot λ_c versus $k_{3\mathrm{f}}$ (k_3) in Fig. 8.1.

According to Fig. 8.1, for $k_{3\mathrm{f}} = 1.5$, $\lambda_c = 0.86$, and for the given coefficients, $\lambda = 1.5625 > \lambda_c$, hence the asymmetric steady states exist, and they are stable. We can also compute (α_a, β_a) using the Mathematica® code given and explore the various relationships between asymmetric steady states and the coefficients.

We compute the numerical solutions of the ODEs given by Eq. (8.5) and Eq. (8.6) with different initial conditions for $[X_L]$ and $[X_D]$. To run the program, there should be a slight difference between the initial conditions, otherwise, symmetric conditions would be the result of computation. We change the initial conditions for three different initial conditions of $[X_L]$ without changing the initial conditions of $[X_D]$, and the time evolution of $[X_L]$ and $[X_D]$ are given in Fig. 8.2. It is shown that smaller the difference between initial values of $[X_L]$ and $[X_D]$, the bifurcation takes longer times. From the equations given in Sect. 8.3, we can compute the asymmetric steady state values which are the same according to Fig. 8.2.

We compute the rate of entropy production per unit volume for each reaction, R1, R2, R3, R4, and R5, and the total rate of entropy production for five reactions. (See Mathematica® code given in Appendix 8.1) Fig. 8.3 shows the rates of entropy production per unit volume for the reactions.

According to Fig. 8.3, R3 is the irreversible reaction which increases the rate of entropy production after the bifurcation point, and all other curves show a drop in the entropy production rate beyond the critical value, the time of which coincides with that of the red and purple curves in Fig. 8.2. R5 is the highest contributing reaction to the entropy production rate, and the total entropy drops after the bifurcation point to a steady state just like rest of the other curves. Since all the curves are in the positive

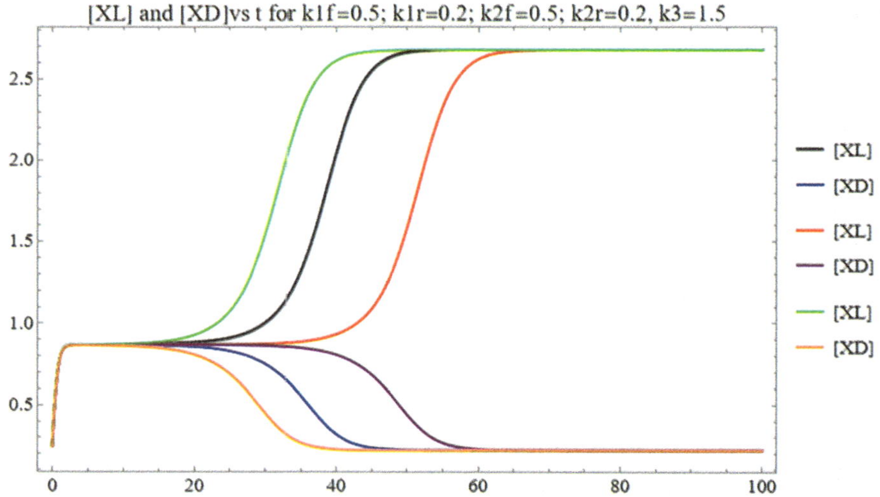

Fig. 8.2 The time evolutions of $[X_L]$ and $[X_D]$ reaching the asymmetric steady states. The graphs in green and orange are solutions for the initial conditions, $[X_L]_0 = 0.251$ and $[X_L]_0 = 0.25$; the graphs in black and blue are the solutions for $[X_L]_0 = 0.2502$ and $[X_L]_0 = 0.25$; and the red and purple curves are for $[X_L]_0 = 0.25001$ and $[X_L]_0 = 0.25$

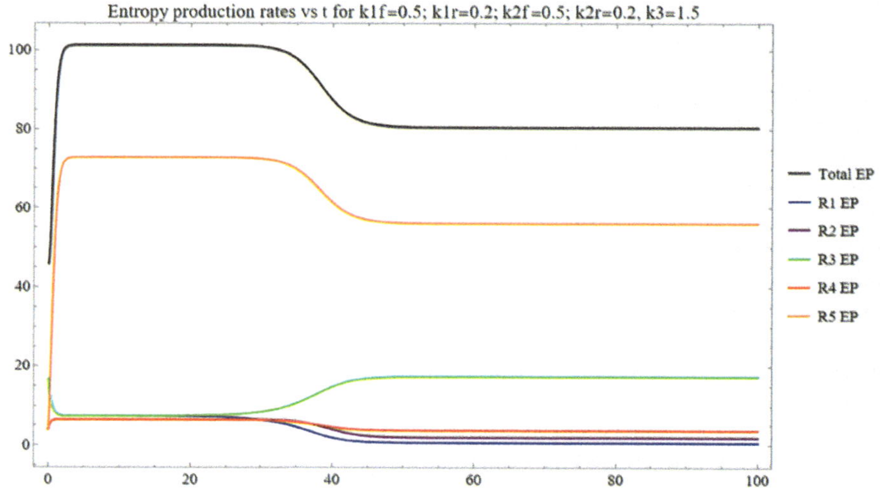

Fig. 8.3 The rate of entropy production per unit volume for the reactions individually and total rate of entropy production based on the coefficient values given and the initial conditions for the ODEs are $[X_L]_0 = 0.25001$ and $[X_L]_0 = 0.25$, corresponding to the red and purple curves in Fig. 8.2

region, the system is stable in the asymmetric state confirming the results we have seen previously.

8.5 A Stochastic KNM (SKNM)

8.5.1 Model Formulation

We formulate a stochastic version of KNM to understand the chirality when the species are reacting within a noisy environment. In Sect. 8.3, we have seen the importance of parameters, i.e., the rate coefficients, $[S]$ and $[T]$ play an important role in asymmetry of $[X_L]$ and $[X_D]$ if the parameters meet specific conditions. In stochastic formulation of KNM, we can choose different ways to incorporate the noise in the system. Keeping with the theme of this monograph, we use the white noise probability space with Wick products to develop the SKNM. An assumption we make here to simply the computations is that we ignore the intrinsic noise in the coefficients, as our experience with other models shows us that the multiplicative noise, a model to impose both extrinsic and intrinsic noises, is sufficient to explore the impact of noise effectively. We could include the noise in multiplicative way in Eqs. (8.6) and (8.7) or in Eqs. (8.17) and (8.18) but it is more intuitive to include the noise in Eqs. (8.6) and (8.7). Another important assumption is that the deterministic ODEs provide the expected values of the SKNM, which can be seen when the expectation is taken across the WSDEs we develop for SKNM.

Let us rewrite the ODEs for $[X_L]$ and $[X_D]$ with the initial values $[X_L] = x_{L0}$ and $[X_D] = x_{D0}$:

$$\frac{d[X_L]}{dt} = k_{1f}[S][T] - k_{1r}[X_L] + k_{2f}[X_L][S][T] - k_{2r}[X_L]^2 - k_{3f}[X_L][X_D] + k_{3r}[P] \tag{8.39}$$

and

$$\frac{d[X_D]}{dt} = k_{1f}[S][T] - k_{1r}[X_D] + k_{2f}[X_D][S][T] - k_{2r}[X_D]^2 - k_{3f}[X_L][X_D] + k_{3r}[P] \tag{8.40}$$

The difference between these ODEs and Eqs. (8.6) and (8.7) is the inclusion of [P] with a negligibly small k_{3r} value. For the notational simplicity, we denote the expectation of $[X_L]$ as $f_1(t)$ and the expectation of $[X_D]$ as $f_2(t)$. We consider [P] as a fixed given value, and $\lambda = [S][T]$, as before a given fixed value. Then we write WSDEs for SKNM as follows:

$$\frac{d[X_L(t,\omega)]}{dt} = k_{1f}\lambda - k_{1r}[X_L] + k_{2f}\lambda[X_L(t,\omega)]$$

$$- k_{2r}[X_L(t,\omega)]^{\diamond 2} - k_{3f}[X_L(t,\omega)]\diamond[X_D(t,\omega)] + k_{3r}[P] + \alpha_1 X_L(t,\omega)\diamond w_1(t,\omega) \tag{8.41}$$

and

$$\frac{d[X_D(t,\omega)]}{dt} = k_{1f}\lambda - k_{1r}[X_D(t,\omega)] + k_{2f}\lambda[X_D(t,\omega)] - k_{2r}[X_D(t,\omega)]^{\diamond 2} - k_{3f}[X_L(t,\omega)]\diamond[X_D(t,\omega)] + k_{3r}[P] + \alpha_2 X_D(t,\omega)\diamond w_2(t,\omega) \tag{8.42}$$

where α_1 and α_2 are the amplitudes or coefficients for the noise terms; $w_1(t,\omega)$ and $w_2(t,\omega)$ are two independent white noise processes. We can simplify the computations by assuming that $\alpha_1 = \alpha_2 = \alpha$ and $w_1(t,\omega) = w_2(t,\omega) = w(t,\omega)$, which is a reasonable assumption given the reactions happen within the same enclosure. If there is a need, we can easily expand the numerical solutions with two independent noise coefficients and white noise processes. However, we can gain important insights with the simplification of WSDEs given above as well.

As in previous chapters, we use the Clark-Ocone formula to write:

$$[X_L(t,\omega)] = f_1(t) + \int_0^t \varphi_1(t,\omega)dW_1 \text{ and} \tag{8.43}$$

$$[X_D(t,\omega)] = f_2(t) + \int_0^t \varphi_2(t,\omega)dW_2 \tag{8.44}$$

where we use two independent Wiener processes, $W_i (i = 1, 2)$, and $\varphi_i (i = 1, 2)$ are the Malliavin derivatives. To simplify the notation, we can define the following stochastic variables:

$$s_1 = \int_0^t \varphi_1(t,\omega)dW_1, \text{ and} \tag{8.45}$$

$$s_2 = \int_0^t \varphi_2(t,\omega)dW_2 \tag{8.46}$$

giving

$$[X_L(t,\omega)] = f_1(t) + s_1, \text{ and} \tag{8.47}$$

$$[X_D(t,\omega)] = f_2(t) + s_2 \tag{8.48}$$

Then we write WSDEs as follows:

$$\begin{aligned}\frac{df_1}{dt}+\frac{ds_1}{dt} &= k_{1f}\lambda - k_{1r}(f_1+s_1)+k_{2f}\lambda(f_1+s_1)\\ &\quad - k_{2r}(f_1+s_1)^{\diamond 2} - k_{3f}(f_1+s_1)\diamond(f_2+s_2)\\ &\quad + k_{3r}[P] + \alpha(f_1+s_1)\diamond w(t,\omega), \text{ and}\end{aligned} \tag{8.49}$$

$$\begin{aligned}\frac{df_2}{dt}+\frac{ds_2}{dt} &= k_{1f}\lambda - k_{1r}(f_2+s_2)+k_{2f}\lambda(f_2+s_2)\\ &\quad - k_{2r}(f_2+s_2)^{\diamond 2} - k_{3f}(f_1+s_1)\diamond(f_2+s_2)\\ &\quad + k_{3r}[P] + \alpha(f_2+s_2)\diamond w(t,\omega).\end{aligned} \tag{8.50}$$

As f_1 and f_2 are deterministic functions, we separate the deterministic ODE from stochastic ODE as we have done before, and simplify the algebra to obtain the following equations for SKNM:

$$\frac{df_1}{dt} = k_{1f}\lambda - k_{1r}f_1 + k_{2f}f_1\lambda - k_{2r}f^2 - k_{3f}f_1f_2 + k_{3r}[P] \tag{8.51}$$

$$\begin{aligned}\frac{ds_1}{dt} &= -k_{1r}s_1 + k_{2f}\lambda s_1 - 2k_{2r}f_1s_1 - k_{2r}s_i^{\diamond 2}\\ &\quad - k_{3f}f_1s_2 - k_{3f}f_2s_1 - k_{3f}s_1\diamond s_2\\ &\quad + \alpha f_1 w + \alpha s_1\diamond w\end{aligned} \tag{8.52}$$

$$\frac{df_2}{dt} = k_{1f}\lambda - k_{1r}f_2 + k_{2f}f_2\lambda - k_{2r}f_2^2 - k_{3f}f_1f_2 + k_{3r}[P]; \text{ and} \tag{8.53}$$

$$\begin{aligned}\frac{ds_2}{dt} &= -k_{1r}s_2 + k_{2f}\lambda s_2 - 2k_{2r}f_2s_2 - k_{2r}s_2^{\diamond 2}\\ &\quad - k_{3f}f_1s_2 - k_{3f}f_2s_1 - k_{3f}s_1\diamond s_2 + \alpha f_2 w + \alpha s_2\diamond w.\end{aligned} \tag{8.54}$$

We give the Mathematica® code for the numerical solution in Appendix 8.2.

8.5.2 Computations with SKNM

We develop the Mathematica® routines that are given in Appendix 8.2 to show the time evolutions of $[X_L]$ and $[X_D]$, which are stochastic dependent variables now, for given amplitude (α) of the noise terms for single realizations at a time. We use the same realization of Wiener increments for comparison purposes. The coefficient values and initial conditions are same as in the deterministic case.

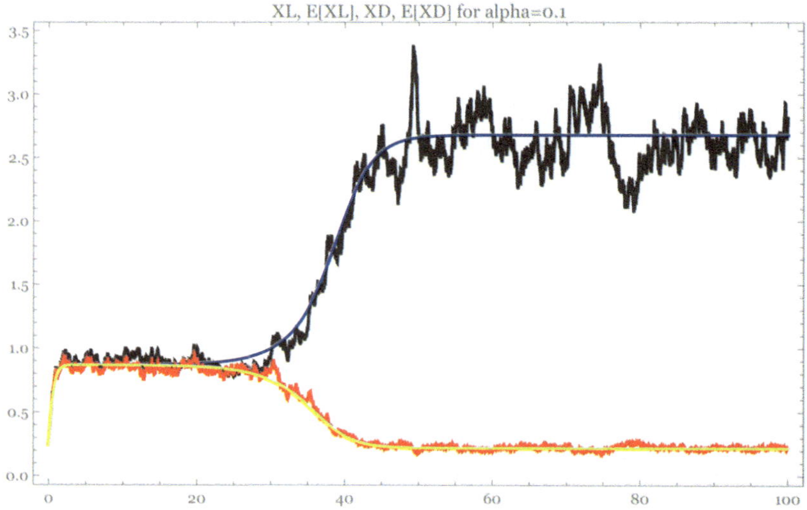

Fig. 8.4 $[X_L]$ and $[X_D]$ realizations overlayed on the respective expectations of $[X_L]$ and $[X_D]$, for $\alpha = 0.1$. $[X_L]$-black; $[X_D]$-red; $E[[X_L]]$-blue; and $E[[X_D]]$-yellow

$$\left(k1\mathrm{f} = 0.5;\ k_{1\mathrm{r}} = 0.2;\ k_{2\mathrm{f}} = 0.5;\ k_{3\mathrm{f}} = k_3 = 1.5;\ k_{3\mathrm{r}} = 10^{-3};\right.$$
$$\left.[S] = 1.25;\ [T] = [S];\ [P] = 0.5;\ [X_L]_0 = 0.25001 \text{ and } [X_L]_0 = 0.25\right)$$

Since $\lambda = 1.5625 > \lambda_c$, the system is stable and SKNM should show stability for moderate values of α. Figure 8.4 shows the realizations of $[X_L]$ and $[X_D]$ for $\alpha = 0.1$ along with the expectations of $[X_L]$ and $[X_D]$. While the realizations follow the expected values, the bifurcation of chiral asymmetry starts around $t = 30$ shifting toward the right.

When we double the amplitude of the noise term to 0.2 (Fig. 8.5), we see higher fluctuations around the expected values in both $[X_L]$ and $[X_D]$realizations, and chiral asymmetry bifurcates at a time greater than 30. This trend continues with increasing α, and higher α values would delay the bifurcation further. Reader can conduct various computational experiments by running the Mathematica® source codes provided in Appendix 8.2.

By moving λ value closer to 0.86, we investigate SKNM behavior when $\alpha = 0.2$ by making $[S] = [T] = 1.0$; we still have the stability but the SKNM behavior changes dramatically. Figure 8.6 shows the realizations of $[X_L]$ and $[X_D]$ for $\lambda = 1.0$.

While the deterministic Chirality asymmetry starts around $t = 80$, the SKNM delays the bifurcation to around $t = 140$, which means that we do not observe the asymmetric chirality soon after deterministic bifurcation point as in the case when $\lambda = 1.56$, and noise can destroy the asymmetric chirality and make it appear to be symmetric in the experiments.

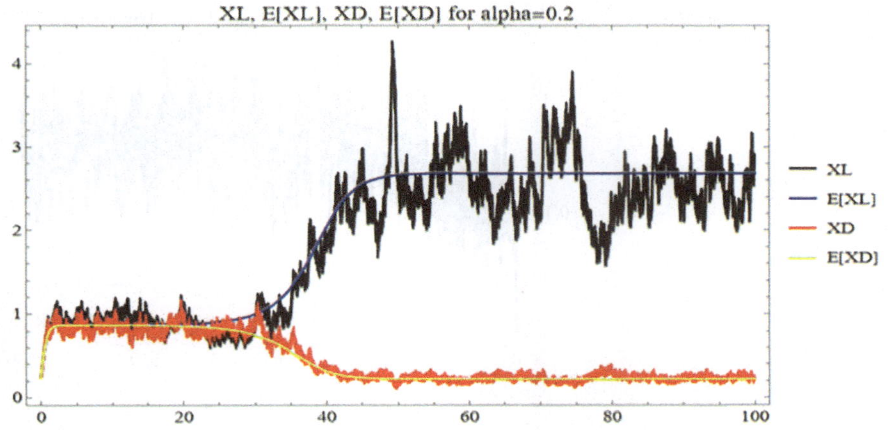

Fig. 8.5 $[X_L]$ and $[X_D]$ realizations overlayed on the respective expectations of $[X_L]$ and $[X_D]$, for $\alpha = 0.2$

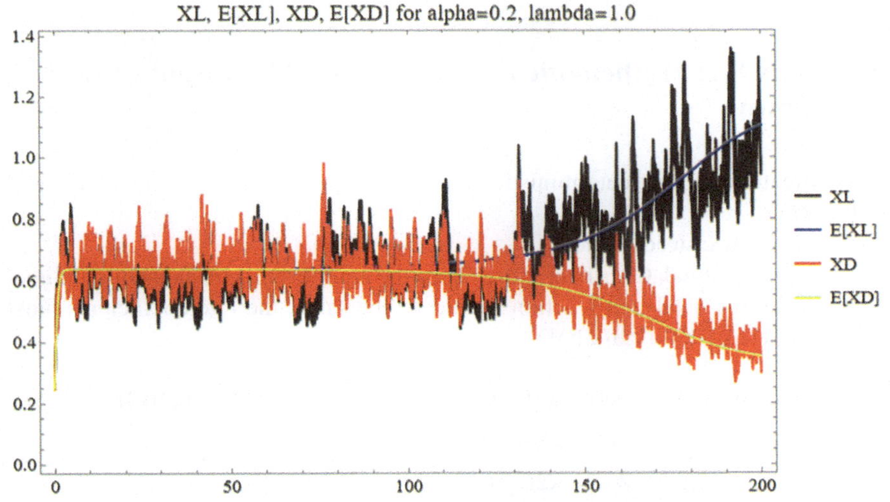

Fig. 8.6 $[X_L]$ and $[X_D]$ realizations overlayed on the respective expectations of $[X_L]$ and $[X_D]$, for $\alpha = 0.2$ and $\lambda = 1.0$

We extend the Mathematica® source code to multiple realizations with multiple Wiener realizations in Appendix 8.3. When we have 30 realizations of $[X_L]$ and $[X_D]$, the noise ratios (NRs) can be computed using the 30 values at each time point as shown in the program. (We have defined the NR previously and see Appendix 8.3 for computations). Figure 8.7 shows the noise ratio realizations of $[X_L]$ and $[X_D]$ when $\alpha = 0.2$ and $\lambda = 1.0$. It shows that both realizations have similar magnitudes across the timeframe indicating that, even though $[X_L]$ and $[X_D]$ are very different in

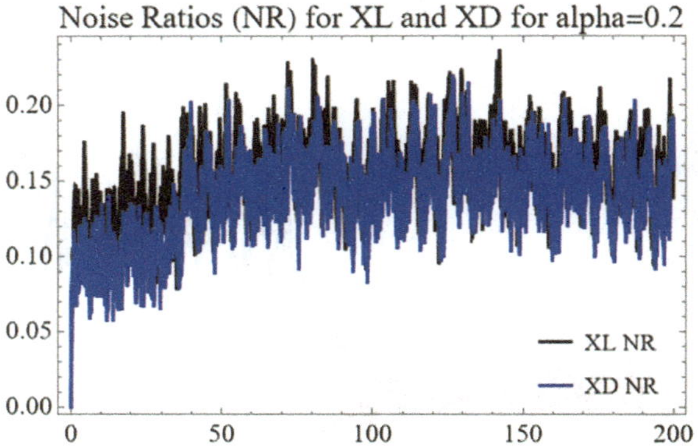

Fig. 8.7 Noise ratios for $[X_L]$ and $[X_D]$ $\alpha = 0.2$ and $\lambda = 1.0$

magnitudes, stochasticity imposed by the model is quite similar for both variables. This shows indirectly that SKNM is a robust model to understand the fluctuations related asymmetric chirality.

Appendix 8.1: Mathematica® Codes for KNM (Outputs Are Not Shown)

(*Investigation of the deterministic model for Chiral symmetry breaking and stability*).

(*Asymmetric steady states*).

(*Compute lambdaC, critical value; since the equilibrium constants of the direct reaction and the catalyzed reaction should be the same, the rate coefficients must adhere to (k1f/k1r) = (k2f/k2r)*)

```
lambdaC[k1f_, k1r_, k2f_, k2r_, k3_] := Module[{s = 2k2fk1r
  + ((4k2r^2k1f)/(k3 − k2r))},
  (s + Sqrt[s^2 − 4(k2f^2)(k1r^2)])/(2k2f^2)]
```

(*Let assume k1f = 0.5; k1r = 0.2; k2f = 0.5; k2r = 0.2; k3 > 0.2, and compute lambdaC–critical lambda for bifurcation. We plot lambdaC versus k3 below*)

```
Plot[lambdaC [0.5, 0.2, 0.5, 0.2, k3], {k3, 0.25, 3.}, PlotRange → Full,

Axes → True, AxesLabel → {Automatic}, Frame → True,

PlotLegends → {λc}, PlotStyle → {Black, Thick},
```

$$\text{PlotLabel} \rightarrow \lambda_c \text{versus k3 for k1f} = 0.5;$$
$$\text{k1r} = 0.2; \text{k2f} = 0.5; \text{k2r} = 0.2,$$

(*Asymmetric steady states*)

$$\text{betaAsy[k1f\_, k1r\_, k2f\_, k2r\_, k3\_, lambda\_]} :$$
$$= (\text{k2f lambda} - \text{k1r})/((2\text{k2r}));$$

(*Alpha value for asymmetric steady state is calculated*).

alphaAsy[k1f_,k1r_,k2f_,k2r_,k3_,lambda_].

: = Sqrt[((k2f lambda- k1r)/((2* k2r)))^2 – ((k1f)/(k3-k2r))];

$$\text{Plot}\big[\text{alphaAsy}\,[0.5, 0.2, 0.5, 0.2, \text{k3}, 1.6], \{\text{k3}, 0.25, 3.\}, \text{Axes} \rightarrow \text{True},$$

$$\text{AxesLabel} \rightarrow \{\text{Automatic}\}, \text{Frame} \rightarrow \text{True},$$

$$\text{PlotLegends} \rightarrow \{\beta_a\}, \text{PlotStyle} \rightarrow \{\text{Black, Thick}\},$$

$$\text{PlotLabel} \rightarrow \beta_a \text{versus k3 for k1f} = 0.5;\ \text{k1r} = 0.2;\ \text{k2f} = 0.5;\ \text{k2r} = 0.2,\ \lambda = 1.6,$$

$$\text{LabelStyle}- > \{\text{FontFamily}$$
$$\rightarrow \text{Times New Roman}, 14, \text{GrayLevel}[0]\}\big]$$

(*Following Kondepudi and Prigogine (2014), the following Mathematica program shows chiral symmetry breaking. As mentioned before the kinetic constants are chosen such that the equilibrium constant, (kf/kr), for direct and catalyzed reactions are equal*).

(*We set k3r to be a very small value *).

$$\text{k1f} = 0.5; \text{k1r} = 0.2; \text{k2f} = 0.5; \text{k2r} = (\text{k1r}/\text{k1f}) * \text{k2f};$$
$$\text{k3f} = 1.5; \text{k3r} = 10^{-3}; S = 1.25; T = S; P = 0.5;$$
$$\text{R1f} := \text{k1f} * S * T; \text{R1r} := \text{k1r} * XL[t];$$
$$\text{R2f} := \text{k2f} * S * T * XL[t]; \text{R2r} := \text{k2r} * (XL[t])^2;$$
$$\text{R3f} := \text{k1f} * S * T; \text{R3r} := \text{k1r} * XD[t];$$
$$\text{R4f} := \text{k2f} * S * T * XD[t]; \text{R4r} := \text{k2r} * (XD[t])^2;$$
$$\text{R5f} := \text{k3f} * XL[t] * XD[t]; \text{R5r} := \text{k3r} * P;$$

(* Initial values of XL and XD are set at steady state values for P = 0.5; we experiment with initial values; there should be a slight difference in initial values for the calculation XL[0] = 0.2502, XD[0] = 0.2500*)

$$\begin{aligned}\text{Soln1} = \text{NDSolve}\big[\{\text{XL}'[t] = \\ = \text{R1f} - \text{R1r} + \text{R2f} - \text{R2r} - \text{R5f} + \text{R5r},\end{aligned}$$

$$\begin{aligned}\text{XD}'[t] == \text{R3f} - \text{R3r} + \text{R4f} - \text{R4r} - \text{R5f} + \text{R5r}, \text{XL}[0] = \\ = 0.2502,\end{aligned}$$

$$\text{XD}[0] == 0.25\}, \{\text{XL}, \text{XD}\}, \{t, 0, 2000\}, \text{MaxSteps}- > 10000\big]$$

$$\text{tmax} = 100;$$

$$\text{Remove}[t]$$

$$\text{Plot}[\text{Evaluate}\,[\{\text{XL}[t], \text{XD}[t]\}/.\text{Soln1}], \{t, 0, \text{tmax}\},$$

$$\text{AxesLabel} \rightarrow \{\text{time}, [\text{XL}]\&\ [\text{XD}]\},$$

$$\text{AxesStyle} \rightarrow \text{Directive}[\text{Black}, 14],$$

$$\text{PlotStyle} \rightarrow \{\{\text{Black}, \text{Thick}\}, \{\text{Blue}, \text{Thick}\}\}, \text{PlotRange} \rightarrow \text{All},$$

$$\text{Axes} \rightarrow \text{True}, \text{AxesLabel} \rightarrow \{\text{Automatic}\}, \text{Frame} \rightarrow \text{True},$$

$$\begin{aligned}\text{PlotLegends} \rightarrow \{[\text{XL}], [\text{XD}]\},\ \text{PlotLabel} \\ \rightarrow [\text{XL}]\,\text{and}\,[\text{XD}]\text{versus}\,t\,\text{for}\,\text{k1f} = 0.5; \\ \text{k1}r = 0.2;\ \text{k2f} = 0.5;\ \text{k2}r = 0.2,\ \text{k3} = 1.5,\end{aligned}$$

$$\text{LabelStyle}- > \{\text{FontFamily} \rightarrow\ \text{Times New Roman}, 14, \text{GrayLevel}[0]\}\big]$$

(*Change the initial conditions–make the difference bigger and smaller and plot the solutions using the similar plot*).

(*Entropy production from Kondepudi and Prigogine (2014)*).

$$\begin{aligned}&\text{RR} = 8.314; \\ &\text{Sig1} := \text{RR} * (\text{R1f} - \text{R1r}) * \text{Log}[\text{R1f}/\text{R1r}]; \\ &\text{Sig2} := \text{RR} * (\text{R2f} - \text{R2r}) * \text{Log}[\text{R2f}/\text{R2r}];\end{aligned}$$

```
Sig3 := RR * (R3f - R3r) * Log[R3f/R3r];
Sig4 := RR * (R4f - R4r) * Log[R4f/R4r];
Sig5 := RR * (R5f - R5r) * Log[R5f/R5r];
Sig Tot := Sig1 + Sig2 + Sig3 + Sig4 + Sig5;

Soln1 = NDSolve[{XL'[t] == R1f - R1r + R2f - R2r - R5f + R5r,

XD'[t] == R3f - R3r + R4f - R4r - R5f + R5r, XL[0] =
= 0.2502,

XD[0] == 0.25}, {XL, XD}, {t, 0, 2000}, MaxSteps- > 10000]

(*Plot the rates of entropy production*)

Plot[Evaluate[{SigTot, Sig1, Sig2, Sig3, Sig4, Sig5}/.Soln1], {t, 0, 100},

Axes → True, Frame → True,

PlotLegends → {Total EP, R1 EP, R2 EP, R3 EP, R4 EP, R5 EP},

PlotLabel
→ Entropy production rates versus t for k1f = 0.5; k1r
= 0.2; k2f = 0.5; k2r = 0.2, k3 = 1.5,

LabelStyle → {FontFamily
→ Times New Roman, 14, GrayLevel[0]}

AxesStyle → Directive[Black, 14],

PlotStyle
→ {{Black, Thick}, {Blue, Thick}, {Purple, Thick}, {Green, Thick},
{Red, Thick}, {Orange, Thick}},
PlotRange- > All]
```

Appendix 8.2: Mathematica® Code for SKNM Single Realizations (Outputs Are Not Shown)

(*Computations for SKNM with one realization at a time *).
(*We use the same parameter values as in Appendix 8.1 *).

```
k1f = 0.5; k1r = 0.20; k2f = 0.5; k2r = (k1r * k2f)/k1f; deltaT = 0.01;
k3f = 1.5;
k3r = 10^-3; S = 1.25;
T = S; lambda = S * T;
XL0 = 0.2502; XD0 = 0.2500; P = 0.5;
```

(*We will assume that same Weiner process would drive both XL and XD*)

```
dBList = Table[N[Random
        [NormalDistribution[0., Sqrt[0.01]]], 32], {10001}];
```

(*Solving the deterministic part *)

```
SolnE = NDSolve[{f1'[t] == k1f * lambda - k1r * f1[t]
        + k2f * lambda * f1[t] - k2r * (f1[t])^2
        - k3f * f1[t] * f2[t] + k3r * P,

f2'[t] == k1f * lambda - k1r * f2[t] + k2f * lambda * f2[t]
        - k2r * (f2[t])^2 - k3f * f1[t] * f2[t]
        + k3r * P, f1[0] == 0.2502, f2[0] == 0.25},
        {f1, f2}, {t, 0, 200}, MaxSteps- > 10000]

Plot[Evaluate[{f1[t], f2[t]}/.SolnE], {t, 0, 100},

AxesLabel → {"time", ["XL]&[XD]"},

AxesStyle → Directive[Black, 14],

PlotStyle → {{Black, Thick}, {Blue, Thick}}, PlotRange → All,

Axes → True, AxesLabel → {Automatic}, Frame → True,
```

```
PlotLegends → {"[XL]"," [XD]"},

PlotLabel
→" [XL] and [XD]versus t for k1f = 0.5; k1r = 0.2; k2f = 0.5; k2r = 0.2, k3 = 1.5",

LabelStyle− > {FontFamily
        → Times New Roman, 14, GrayLevel[0]}]

f1f2List = Table[Evaluate [{f1[t], f2[t]}/.SolnE], {t, 0, 100, 0.01};

f1List = Table[f1f2 List][[i, 1]][[1]], {i, 1, 10001}];

f2List = Table[f1f2List [[i, 1]][[2]], {i, 1, 10001}];

f1tList = Table[{(i − 1)0.01, f1List[[i]]}, {i, 1, 10001}];

f2tList = Table[{(i − 1)0.01, f2List[[i]]}, {i, 1, 10001}];

(*Solving the stochastic part *)

alpha = 0.1;

S1List = Table[0., {i, 1, 10001}];

S2List = Table[0., {i, 1, 10001}];

Do[S1List [[i + 1]]
    = S1List[[i]] + (−k1r ∗ S1List [[i]] + k2f
    ∗ lambda ∗ S1List [[i]] − 2. ∗ k2r ∗ f1 List[[i]]
    ∗ S1List[[i]] − k2r ∗ (S1List[[i]])^2 − k2f
    ∗ f1List[[i]] ∗ S2List[[i]] − k3f ∗ f2List[[i]]
    ∗S1List[[i]] − k3f ∗ S1List[[i]] ∗ S2List[[i]])
    ∗ 0.01 + alpha ∗ f1List [[i]] ∗ dBList[[i]] + alpha
    ∗ S1List[[i]] ∗ dBList[[i]]; S2List[[i + 1]]
    = S2List[[i]] + (−k1r ∗ S2 List[[i]]) + k2f
    ∗ lambda ∗ S2List [[i]] − 2. ∗ k2r ∗ f2 List[[i]]
    ∗ S2List[[i]] − k2r ∗ (S2 List[[i]]) ∧ 2 − k3f
```

```
* f1List[[i]] * S2List[[i]] - k3f * f2List[[i]]
*S1List[[i]] - k3f * S1List[[i]] * S2List[[i]])
* 0.01 + alpha*f2List[[i]] * dBList[[i]] + alpha
*S2List[[i]]*dBList[[i]], {i, 1, 10000}];

XLList = Table[0., {i, 1, 10001}];
XLList = f1List + S1List;

XDList = Table[0., {i, 1, 10001}];
XDList = f2List + S2List;

XLtList = Table[{(i - 1)0.01, XLList[[i]]}, {i, 1, 10001}];
XDtList = Table[{(i - 1)0.01, XDList[[i]]}, {i, 1, 10001}];

ListLinePlot[{XLtList, f1tList, XDtList, f2tList}, Frame -> True,

Axes -> True, PlotLabel ->" XL, E[XL], XD, E[XD]for alpha = 0.1",

AxesLabel -> {t},

PlotStyle -> {{Black, Thick}, {Blue, Thick}, {Red, Thick}, {Yellow, Thick}},

PlotRange -> All, PlotLegends -> {XL, E[XL], XD, E[XD]},

LabelStyle- > {FontFamily
    ->" Times New Roman", 14, GrayLevel[0]}]
```

Appendix 8.3: Mathematica® for SKNM Multiple Realizations

```
(*This is SKNM with multiple realizations *)

k1f = 0.5; alpha = 0.2; k1r = 0.2; k2f = 0.5;

k2r = (k1r * k2f)/k1f;
```

$$\text{deltaT} = 0.01;\ \text{k3f} = 1.5;\ \text{k3r} = 10^{-3};\ S = 1.25;\ T = S;$$

$$\text{lambda} = S * T;$$

$$\text{XL0} = 0.2502;\ \text{XD0} = 0.25;\ P = 0.5;$$

(*We will assume that same Weiner process realization would drive both XL and XD*)

```
dBList
 = Table[N [Random [NormalDistribution
   [0., Sqrt[deltaT]]], 32], {20001}, 30];
dBListT = Transpose[dBList];
```

```
SolnE = NDSolve[
{f1'[t] == k1f * lambda - k1r * f1[t] + k2f * lambda * f1[t]
-k2r * (f1[t])^2 - k3f * f1[t] * f2[t] + k3r * P,
 f2'[t] == k1f * lambda - k1r * f2[t] + k2f * lambda * f2[t]
      -k2r * (f2[t])^2 - k3f * f1[t] * f2[t] + k3r * P,
  f1[0] == 0.2502, f2[0] == 0.25}, {f1, f2}, {t, 0, 2000},
              MaxSteps- > 10000]
```

```
f1f2List = Table[Evaluate [{f1[t], f2[t]}/.SolnE], {t, 0, 200, 0.01}];

f1List = Table[f1f2List [[i, 1]][[1]], {i, 1, 20001}];

f2List = Table[f1f2List [[i, 1]][[2]], {i, 1, 20001}];

f1Matrix = Table[f1List[[i]], {i, 1, 20001}, {j, 1, 30}];

f1MatrixT = Transpose[f1Matrix];

f2Matrix = Table[f2List[[i]], {i, 1, 20001}, {j, 1, 30}];
```

```
f2MatrixT = Transpose[f2Matrix];

S1List = Table[0., {i, 1, 20001}, {j, 1, 30}];

S1ListT = Transpose[S1List];

S2List = Table[0., {i, 1, 20001}, {i, 1, 30}];

S2ListT = Transpose[S2List];

Do[S1ListT[[j, i + 1]]
    = S1ListT[[j, i]] + (-k1r * S1ListT[[j, i]] + k2f
    * lambda * S1ListT[[j, i]] - 2. * k2r
    *f1MatrixT[[j, i]] * S1ListT[[j, i]] - k2r
    * (S1ListT[[j, i]]) ∧ 2 - k2f * f1MatrixT[[j, i]]
    * S2ListT[[j, i]]) * deltaT + alpha
    *f1MatrixT[[j, i]] * dBListT[[j, i]] + alpha
    * S1ListT[[j, i]] * dBListT[[j, i]];

 S2ListT[[j, i + 1]]
    = S2ListT[[j, i]] + (-k1r* S2ListT[[j, i]] + k2f
    * lambda * S2ListT[[j, i]] - 2. * k2r
    *f2MatrixT[[j, i]]*S2ListT[[j, i]] - k2r
    *(S2ListT[[j, i]])^2 - k3f * f1MatrixT[[j, i]]
    * S1ListT[[j, i]] - k3f * S1ListT[[j, i]]
    * S2ListT [[j, i]]) * deltaT + alpha
    * S2ListT[[j, i]]
    * dBListT[[j, i]], {i, 1, 20000}, {j, 1, 30}]

XLList = Table[0., {i, 1, 20001}, {j, 1, 30}];
XLListT = Transpose[XLList];

XLListT = f1MatrixT + S1ListT;
```

XDList = Table[0., {i, 1, 20001}, {j, 1, 30}];
XDListT = Transpose[XDList];
XDListT = f2MatrixT + S2ListT;

meanXLListT = Mean[XLListT];
varXLListT = Variance[XLListT];

noiseRatioXL
= Table[(Sqrt[varXLListT[[i]]]/meanXLListT[[i]]), {i, 1, 20001}];

noiseRatioXLt = Table[{(i − 1)deltaT, noiseRatioXL[[i]]}, {i, 1, 20001}];

meanXDListT = Mean[XDListT];
varXDListT = Variance[XDListT];

noiseRatioXD
= Table[(Sqrt[varXDListT[[i]]]/meanXDListT[[i]]), {i, 1, 20001}];

noiseRationXDt
= Table[{(i − 1)deltaT, noiseRatioXD[[i]]}, {i, 1, 20001}];

ListLinePlot[{noiseRatioXLt, noiseRationXDt}, Frame → True,

Axes → True, PlotLabel
→ Noise Ratios (NR) for XL and XD for alpha = 0.2,

AxesLabel → {t}, PlotStyle → {{Black, Thick}, {Blue, Thick}},

PlotRange → All, PlotLegends → {XL NR, XD NR},

LabelStyle− > {FontFamily
→″ TimesNewRoman″, 14, GrayLevel[0]}]

References

Colwell, J. A. (2022). A necessary and sufficient condition in the model of Kondepudi and Nelson for the breaking of chiral symmetry. *Dynamics*, *2*(3). Available at: https://doi.org/10.3390/dynamics2030016.

Prigogine, I., & Stengers, I. (2017). *Order out of Chaos Man's new dialogue with nature*. Verso.

Kondepudi, D., & Prigogine, I. (2014). *Modern thermodynamics: From heat engines to dissipative structures* 2nd edn. Available at: https://doi.org/10.1002/9781118698723.

Kondepudi, D. K., & Nelson, G. W. (1983). Chiral symmetry breaking in nonequilibrium systems. *Physical Review Letters*, *50*(14). Available at: https://doi.org/10.1103/PhysRevLett.50.1023.

Kondepudi, D. K., & Nelson, G. W. (1984a). Chiral symmetry breaking in nonequilibrium chemical systems: Time scales for chiral selection. *Physics Letters A*, *106*(4). Available at: https://doi.org/10.1016/0375-9601(84)90320-7.

Kondepudi, D. K., & Nelson, G. W. (1984b). Chiral-symmetry-breaking states and their sensitivity in nonequilibrium chemical systems. *Physica A: Statistical Mechanics and its Applications*, *125*(2–3). Available at: https://doi.org/10.1016/0378-4371(84)90065-7.

Kondepudi, D. K., & Nelson, G. W. (1985). Weak neutral currents and the origin of biomolecular chirality. *Nature*, *314*(6010). Available at: https://doi.org/10.1038/314438a0.

Index

D. Kulasiri, *Stochastic Differential Equations for Chemical Transformations in White Noise Probability Space*, https://doi.org/10.1007/978-981-97-9392-1

The manufacturer's authorised representative in the EU is Springer Nature Customer Service Centre GmbH, Europaplatz 3, 69115 Heidelberg, Germany. If you have any concerns regarding our products, please contact ProductSafety@springernature.com

Printed and bound by CPI Group (UK) Ltd, Croydon, CR0 4YY
01/10/2026
02225916-0014